Cone penetration testing

CIRIA, the Construction Industry Research and Information Association, is an independent non-profit-distributing body which initiates and manages research and information projects on behalf of its members. CIRIA projects relate to all aspects of design, construction, management, and performance of buildings and civil engineering works. Details of other CIRIA publications, and membership subscription rates, are available from CIRIA at the address below.

This CIRIA Ground Engineering Report was written by Dr A. C. Meigh under contract between CIRIA and Woodward-Clyde Consultants, and with the guidance of the project steering group:

B. A. Leach	Allott & Lomax
A. P. Butcher	Building Research Establishment
A. Chalmers	Cementation Piling & Foundations Ltd
J. Daley	Cementation Piling & Foundations Ltd
H. Erwig	Fugro Ltd
J. M. Head	CIRIA (now Sir Alexander Gibb & Partners)
M. B. Jamiolkowski	Politecnico di Torino
S. Thorburn	Thorburn Associates
T. R. M. Wakeling	Delft Geotechnics UK Ltd

The project was funded by:
Department of the Environment
Foundation Engineering Ltd
Fugro Ltd
Soil Mechanics Ltd
Wimpey Laboratories Ltd

CIRIA's research manager for ground engineering is F. M. Jardine

CIRIA
6 Storey's Gate
London SW1P 3AU
Tel. 01–222–8891
Fax. 01–222 1708

CIRIA Ground Engineering Report: In-situ Testing

Cone penetration testing

methods and interpretation

A. C. Meigh, OBE, DSc(Eng), FCGI CEng FICE, MASCE, FGS
Woodward-Clyde Consultants

Construction Industry Research and Information Association

CIRIA

Butterworths
London Boston Durban
Singapore Sydney Toronto Wellington

First published 1987
© **CIRIA, 1987.**

British Library Cataloguing in Publication Data

Meigh, A. C.
   Cone penetration testing : methods and
   interpretation.——(CIRIA ground engineering
   report: in-situ testing).
   1. Soils——Testing——Laboratory manuals
   I. Title   II. Construction
   Industry Research and Information
   Association   IV. Series
   624.1'5136   TA710.5

   ISBN 0–408–02446–1

Library of Congress Cataloging-in-Publication Data

Meigh, A. C.
   Cone penetration testing.
   (CIRIA ground engineering report. In-situ testing)
   Bibliography: p.
   Includes index.
   1. Soil penetration test. I. Title. II. Series.
TA710.5.M43 1987   624.1'5136'0287   87–18283

   ISBN 0–408–02446–1 (Butterworths)

Filmset by Latimer Trend & Company Ltd, Plymouth
Printed and bound in Great Britain by Adlard & Son Ltd, Letchworth, Hertfordshire

Contents

Introduction

This Report provides guidance on the use of the cone penetration test (CPT), and on the interpretation of test results and their use in design. Part 1 covers the CPT as at present established. Part 2 deals with pore-pressure sounding (PPS) and the cone penetration test with pore-pressure measurement (CPTU). These are new techniques, still under development, which promise to increase the effectiveness of penetration testing. The combination of the cone penetrometer with other devices is the subject of Appendix B.

The CPT has not so far been used as widely (and as often) as it might be, and it is hoped that this Report will encourage its use. At the same time, some words of warning are required. Correlations of CPT results with engineering parameters and performance presented herein should be viewed with the appropriate degree of caution. This varies from one correlation to another, and comments on this aspect are included in the text. The value of site specific or local correlations is emphasised. The use of the CPT with pore-pressure measurement should be regarded as experimental at this stage.

The practice described is that adopted in the United Kingdom, which closely follows European practice. Reference is also made to North American methods, particularly in respect of interpretation of results. Although cones for penetration testing are available in various sizes, the only cones considered are, with minor exceptions, those with a 60° point angle and a diameter of 35.7 mm (cross-sectional area 1000 mm^2).

Synopses are provided of Sections 5, 6, 8, 9 and 12. These are intended as aides-memoires, and should not be used without prior reference to or familiarity with the preceding text.

Notation

a	attraction (kN/m^2)
A_b	pile base area (m^2)
A_c	projected area of the cone (mm^2)
A_g	area of groove at base of cone (mm^2)
A_s	surface area of friction sleeve (mm^2)
B	width of foundation (m)
B_q	pore-pressure ratio
c_h	coefficient of horizontal consolidation (m^2/s)
c_u	undrained shear strength (kN/m^2)
c_v	coefficient of (vertical) consolidation (m^2/s)
C	coefficient of compressibility (Terzaghi-Buisman)
C_1	correction for depth of embedment (Schmertmann)
C_2	creep correction (Schmertmann)
C_c	compression index
C_s	swelling index
C_α	coefficient of secondary compression
d	pile diameter (m)
d_b	diameter of pile base (m)
D	depth (m)
D_r	relative density (%)
D_{50}	particle size (mm) corresponding to 50% passing (also D_{10}, D_{60})
e_o	initial void ratio
E	modulus of linear deformation (Young's modulus) (MN/m^2)
E'	effective stress Young's modulus (MN/m^2)
E_u	undrained Young's modulus (MN/m^2)
E_v'	vertical effective stress Young's modulus (MN/m^2)
E_{25}	drained secant modulus at 25% of failure stress (MN/m^2) (also E_{50})
f_s	local side friction (MN/m^2)
F	factor of safety
F_ω	correction factor for pile taper
G	modulus of shear deformation (shear modulus) (MN/m^2)
h_c	height of cone (mm)
H	layer thickness (m)

H_o	thickness of cohesive overburden layer (m)
H_w	depth to groundwater level (m)
I_L	liquidity index
I_P	plasticity index (%)
I_z	strain influence factor
I_{zp}	maximum strain influence factor
k	coefficient of permeability (m/s)
k_h	coefficient of horizontal permeability (m/s)
k_s	correction factor (sands) in pile shaft resistance
k_y	coefficient of vertical permeability (m/s)
K_b	horizontal stress index
K_o	coefficient of earth pressure at rest
K_q	cone resistance correction factor for test chamber size
K_S	coefficient of lateral pressure on pile shaft
l	depth to considered side friction value (m)
L	embedded length of pile (m)
L	length of foundation (m)
m_h	horizontal coefficient of volume change (m²/MN)
m_v	(vertical) coefficient of volume change (m²/MN)
M	constrained modulus (MN/²)
M_o	initial tangent constrained modulus (MN/m²)
n	number of layers
N	SPT blow count
N_c, N_q	bearing capacity factors
N_k	cone factor
N_k^*	corrected cone factor (Bjerrum correction)
p	applied loading (kN/m²)
p_n	net applied loading (kN/m²)
p_L	pressuremeter limit pressure (MN/m²)
p_o	pressure to begin to move membrane (Marchetti dilatometer) (kN/m²)
q_b	ultimate pile end-bearing pressure (kN/m²)
q_c	cone resistance (MN/m²)
$\bar{q}_c$	average cone resistance (MN/m²)
q_c'	effective cone resistance (MN/m²)
q_{co}	(a function of earthquake shaking intensity) used in determining q_{crit} (MN/m²)
q_{crit}	cone resistance below which liquefaction is likely (MN/m²)
q_p	composite value of q_c (MN/m²)
q_s	unit ultimate shaft resistance (kN/m²)
q_T	corrected cone resistance (MN/m²)
Q	ultimate pile bearing capacity (kN)
Q_b	ultimate pile-end bearing capacity (kN)
Q_S	ultimate pile shaft resistance (kN)
Q_T	ultimate pile tension load capacity (kN)
r	number of rows of piles

R	radius of cone tip
R_b	pile bearing capacity reduction factor
R_f	friction ratio (%)
s	settlement (mm)
s_1	settlement of a single pile (mm)
s_g	settlement of a pile group (mm)
s_i	immediate settlement (mm)
s_u	undrained shear strength from vane test (kN/m^2)
s_u^*	corrected undrained shear strength from vane test (Bjerrum correction) (kN/m^2)
S_1	pile shaft resistance coefficient (based on f_s)
S_2	pile shaft resistance coefficient (based on q_c)
S_3	Nottingham's correction factor for additional friction along tapered and step-tapered piles
S_t	sensitivity to remoulding
t	time from load application (years)
t_{50}	time to 50% consolidation (s)
T_{50}	time factor, 50% consolidation
u	pore pressure (MN/m^2)
u_o	equilibrium pore pressure (MN/m^2)
u_{sh}	pore pressure at upper end of friction sleeve (MN/m^2)
u_T	total pore pressure (MN/m^2)
w_L	liquid limit (%)
w_P	plastic limit (%)
W	weight of pile (kN)
x	coefficient in Schmertmann settlement calculation
Y	ratio of pile spacing to pile diameter
Z	depth (m)
Z_c	critical depth of pile (m)
α	modulus coefficient
α_{Eu}	undrained Young's modulus coefficient
$\alpha_{E'_v}$	vertical effective stress Young's modulus coefficient
α_M	constrained modulus coefficient
γ	unit weight (kN/m^3)
γ_w	unit weight of water (kN/m^3)
δ	friction angle, soil to cone
Δ_p	net foundation pressure (Schmertmann) (kN/m^2)
Δ_u	change in pore pressure (MN/m^2)
λ	area correction factor
μ	Bjerrum's correction factor to vane shear strength
ν	Poisson's ratio, total stress
ν'	Poisson's ratio, effective stress
σ'	effective stress (kN/m^2)
σ'_{ho}	effective horizontal stress (kN/m^2)
σ'_p	pre-consolidation pressure (kN/m^2)

σ'_v effective vertical stress (kN/m²)

σ'_{vb} effective vertical stress at pile toe level (kN/m²)

σ'_{vc} effective vertical stress at critical depth (kN/m²)

σ_{vo} overburden pressure (total vertical stress) (kN/m²)

σ'_{vo} effective overburden pressure (effective vertical stress) (kN/m²)

σ'_{vp} effective vertical stress at level of I_{zp} (Schmertmann) (kN/m²)

τ shear stress (kN/m²)

τ_f shear stress at failure (kN/m²)

ϕ angle of shearing resistance (degrees)

ϕ' effective angle of shearing resistance (degrees)

ϕ'_a angle of friction between pile and soil (degrees)

ϕ'_b angle of shearing resistance of sand below pile toe (degrees)

ϕ'_s angle of shearing resistance of sand around pile after installation (degrees)

ω reduction factor applied to end bearing capacity of bored pile in stiff, fissured clay

Glossary

Cone

The part of the penetrometer tip on which the end bearing is developed.

In the *simple cone*, the length of the cylindrical prolongation above the conical part is generally equal to, or less than, the diameter of the cone.

In the *mantle cone*, the cone is prolonged with a gently tapered sleeve, called the mantle, which has a length greater than the diameter of the cone base.

Cone resistance, q_c

The total force acting on the cone, Q_c, divided by the projected area of the cone, A_c.

$$q_c = Q_c/A_c$$

Friction ratio, R_f

The ratio, expressed as a percentage, of the local side friction, f_s, to the cone resistance, q_c, both measured at the same depth.

$$R_f = (f_s/q_c).100$$

Friction reducer

A narrow local protuberance outside the push-rod surface, placed at a certain distance above the penetrometer tip, and provided to reduce the total friction on the push rods.

Friction sleeve

The section of the penetrometer tip upon which local side friction is measured.

Inner rods

Solid rods which slide inside the push rods to extend the tip of a mechanical penetrometer.

Local side friction, f_s

The total frictional force acting on the friction sleeve, Q_f, divided by its surface area, A_s.

$$f_s = Q_f/A_s$$

Penetrometer

An apparatus consisting of a series of cylindrical rods with a terminal body called the *penetrometer tip,* and with the devices for measurement of cone resistance, local side friction and/or total resistance.

The *electric penetrometer* uses electrical devices, such as strain gauges, built into the tip, for measurement.

The *mechanical penetrometer* uses a set of inner rods to operate the penetrometer tip.

Hydraulic or pneumatic penetrometers use hydraulic or pneumatic devices built into the tip.

Push rods

The thick-walled tubes or rods used for advancing the penetrometer tip and to guide and shield part of the measuring system (sometimes called 'sounding tubes').

Reference test penetrometer tip (or reference tip) (R)

A penetrometer tip conforming to reference test requirements.

Thrust machine (rig)

The equipment which pushes the penetrometer tip and rods into the ground.

Part I

The cone penetration test

1 General

In the CPT, a cone on the end of a series of rods is pushed into the ground at a constant rate, and continuous or intermittent measurements are made of resistance to penetration of the cone. If required, measurements are also made of either the combined resistance to penetration of the cone and outer surface of the rods or the resistance of a surface sleeve.

1.1 Historical outline

Probing with rods through weak soils to locate a firmer stratum has been practised for a very long time. Earliest versions of sounding were developed in about 1917 by the Swedish State Railways, then by the Danish Railways around 1927. It was not until 1934, in the Netherlands, that the CPT was introduced in a form recognisable today (Barentsen, 1936). The method has variously been called the Static Penetration Test, Quasi-static Penetration Test, Dutch Sounding Test and Dutch Deep Sounding, and the term 'sounding' is still frequently used. The method was particularly suited to the Netherlands where the ground consists mainly of deltaic deposits, and it was first used as a means of determining the ultimate bearing capacity of driven piles founded in sand.

Initially, the apparatus consisted of a simple cone (Appendix D, Figure 8). Load on the cone was measured as the cone was advanced ahead of the outer tubes, and the total load was measured as the cone and outer tubes were then advanced together. This was soon followed by the 'mantle cone' (Appendix D, Figure 5), designed to prevent soil particles from entering the space between the cone and the push rods (Vermeiden, 1948). A friction sleeve to measure local skin friction over a short length above the cone was then introduced in Indonesia (Begemann, 1953) (Appendix D, Figure 6). The electric penetrometer, in which the cone resistance is measured by transducers mounted immediately above the cone and with local skin friction similarly measured, was first introduced in 1948 (Geuze, 1953), but came into general use in the late 1960s.

The CPT was introduced into the United Kingdom in the 1950s, but it has only been used on a small proportion of ground investigations. Its use has now spread throughout Europe, to the USA and to many other parts of the world.

3

1.2   The role of the CPT

The CPT has three main applications:

1. to determine the soil profile and identify the soils present
2. to interpolate ground conditions between control boreholes
3. to evaluate the engineering parameters of the soils and to assess bearing capacity and settlement.

In this third role, in relation to certain problems, the evaluation is essentially preliminary in nature, to be supplemented by borings and by other tests, either *in situ* or in the laboratory. In this respect, the CPT provides guidance on the nature of such additional testing, and helps to determine the positions and levels at which *in-situ* tests or sampling should be undertaken. Where the geology is fairly uniform and predictions based on CPT results have been extensively correlated with building performance (as in parts of Belgium and the Netherlands), the CPT can be used alone in investigation for building foundations.

Even in these circumstances, CPTs may be accompanied by, or followed by, borings for one or more of the following reasons:

1. to assist where there is difficulty in interpretation of the CPT results
2. to further investigate layers with relatively low cone resistance
3. to explore below the maximum depth attainable by CPT
4. if the project involves excavation, where samples may be required for laboratory testing and knowledge of groundwater levels and permeability is needed.

The depth of penetration is controlled by the ground conditions. Light rigs (see Section 2.1.1) penetrate clays and silts of firm consistency, and they achieve some penetration into loose and medium dense sands. The heavy rig (Section 2.1.3) penetrates stiff clays and loose gravels, but penetration into dense gravels is small. Cobbles and boulders can prevent further progress, and penetration into rock is limited to the weathered upper surface.

Identification of soils (see Section 4.2) is achieved by means of empirical correlations between soil type and the ratio of local side friction to cone resistance (skin friction ratio) considered in relation to the cone resistance. Assessment of engineering parameters (Sections 5 and 6) is also based on empirical correlations. Alternatively, the test results can be directly used to estimate bearing capacity and settlements, again on an empirical basis.

The CPT has two main advantages over the usual combination of boring, sampling and standard penetration testing:

1. It provides a continuous, or virtually continuous, albeit indirect, record of ground conditions.
2. It avoids the disturbance of the ground associated with boring and sampling, particularly that which occurs with the Standard Penetration Test (SPT).

Furthermore, the disturbance resulting from the advancement of the cone is consistent between one test and another. Cost savings can also be made by using the CPT.

1.3 Standardisation

There is no British Standard for the CPT, although one is in preparation. The Sub-committee for the Penetration Test for Use in Europe, a sub-committee of the International Society of Soil Mechanics and Foundation Engineering (ISSMFE), has made recommendations for standardisation of the CPT. These are given in their entirety as Appendix D. This European Recommended Standard (ERS) is in general use in the United Kingdom, and is referred to in BS 5930: 1981, Site investigations.

The ERS covers the geometry of the penetrometer tip; testing procedure; precision of measurements; precautions, checks and verifications; special features; and reporting of results (Appendix D, 1 to 9). In the body of the report presenting the ERS, it was recognised that there would be continuing use of penetrometers which do not

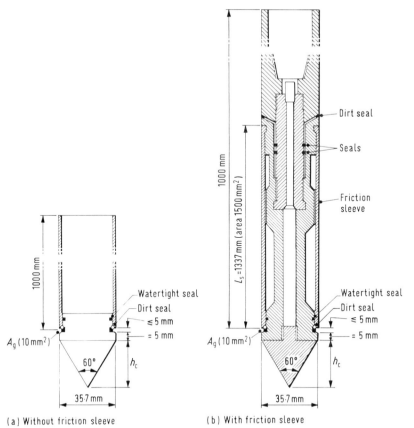

(a) Without friction sleeve   (b) With friction sleeve

Figure 1 An electric penetrometer tip conforming to reference test requirements. For manufacturing and operating tolerances for cone height see p. 128

conform to the standard tip geometry. Provision was therefore made to allow for this (Appendix D, 10). The ERS requires that any deviation from the standard tip geometry and test procedures should be stated when presenting CPT results.

ISSMFE are now working to produce an international reference test procedure (IRTP) for the cone penetrometer. The Technical Committee responsible has been working since 1982, and it has formulated proposals which will shortly be issued for international comment and discussion at a special conference in 1988. The Committee is charged with producing the final IRTP for ratification by ISSMFE in 1989. It is expected that this will define the geometry of a 'reference test penetrometer', which will have the dimensions and features described as 'standard' in the ERS (Appendix D, 3). An example of a penetrometer tip conforming to reference test requirements is shown in Figure 1. ERS describes a tip conforming to these requirements as Standard and ascribes to it the symbol (S). However, in this Report, in anticipation of the International Reference Test Procedure, such a tip is described as a 'reference test penetrometer tip', or 'reference tip', with the symbol (R).

ISSMFE is also expected to recommend procedures for the performance of the cone penetrometer test, generally as set out in the ERS with, in particular, continuous penetration at a standard rate of $20 \pm 5\,mm/s$.

Considerable benefit will derive from standardisation of equipment and procedures (see Sections 2.2.1 and 3.3), and it is strongly recommended that, until the IRTP is published, the ERS should be followed in the United Kingdom, including the adoption of the reference tip (R).

2 Equipment

2.1 CPT apparatus

The CPT apparatus consists of a thrust machine and reaction system (rig), and a penetrometer, including measurement and recording equipment. Machines in the United Kingdom generally have a thrust in the range from 20 to 200 kN. They are discussed below under three categories: light, medium, and heavy. Hand-held and lightweight 'suitcase' apparatus is also available.

A list of European manufacturers is given in Appendix A.

2.1.1 Light rigs

These rigs, usually of 20- or 25-kN capacity, are mainly used for exploration of weak upper layers. They are often hand operated through a chain drive, and they are light enough to be man-handled where access is difficult, although they are sometimes mounted on a trailer or vehicle.

The penetrometer used is commonly fitted with a mechanical mantle cone (see Section 2.2.2) without a friction sleeve. Where cone resistance only is required (often the case with such machines), push rods smaller in diameter than the cone base are used, except for a 1-m length at the top and another immediately above the cone. Penetration is limited to a short distance into medium-dense sands or stiff clays. Load is indicated by a pressure gauge from a hydraulic load cell, or by a proving ring and dial gauge. These rigs can be fitted to borehole casings and used in conventional boreholes as an alternative to SPTs. After refusal is met, the hole is cleaned out and another test may be started.

2.1.2 Medium rigs

These rigs are commonly of 100-kN, but sometimes only of 50-kN thrust capacity. They are usually integral with a trailer mounting (Figure 2) with reaction provided by four- or six-screw picket anchors. However, they can be fitted to a trailer ballasted with kentledge or to a truck, tractor, or other vehicle either of sufficient weight or anchored by screw pickets.

Some earlier 100-kN rigs were hand operated, but few of these remain in use.

Figure 2 Trailer-mounted medium penetrometer rig (courtesy Laboratorium Voor Grondmechanica, Delft)

Penetration is usually achieved by means of a hydraulic jacking system, although rigs with low-friction screws are also in use. Sometimes, the pull-down of a drilling rig is utilised. Reasonable penetration can be obtained in stiff clays and medium-dense sands, perhaps to 20 m, but penetration into dense sands is limited to a few metres, and it is negligible in very dense sands and gravels.

Frequently, a mechanical cone is used, with a friction sleeve, but electrical tips are also used with this size of machine.

The depth of penetration can be increased by the use of a 'friction reducer' or by the technique known as static-dynamic sounding (Sherwood and Child, 1974; Amar, 1974).

Pre-boring, or pre-driving of a heavy cone-ended tube, can also be used, and there are various 'down-the-hole' techniques mainly used in seabed exploration (see Section 2.3).

2.1.3 Heavy rigs

Heavy rigs usually have a thrust capacity in the range 175 to 200kN, which is considered to be the maximum practicable load to avoid buckling of the push rods, although a 300-kN machine is now available. A heavy rig is frequently fitted with an electric cone and friction sleeve (see Section 2.2), but mechanical tips are also used. As with the medium rig, it can be mounted on a lorry, tractor, or trailer, but more often it is fitted to a purpose-built vehicle (Figure 3) with an enclosed area which houses the penetrometer and the associated measuring devices (see Section 3.2). The vehicle can be a truck, usually of the double rear axle rear-and-front-wheel-drive type, or a track-mounted (Figure 4) or 'all-terrain' tractor. The vehicle is ballasted to provide the reaction required, or the weight of the vehicle and ballast is supplemented by screw anchors. A wheeled vehicle is lifted by a number of hydraulic jacks to limit movement during testing.

The power for penetration is usually obtained by a take-off from the main engine, acting on the rods through a thrust head or a hydraulic clamping device.

Figure 3 Heavy penetrometer rig enclosed in a purpose-built vehicle (courtesy Fugro Ltd)

Figure 4 Track-mounted penetrometer (courtesy Fugro Ltd)

Some heavy rigs utilise both an electric penetrometer and a mechanical penetrometer (Bruzzi and Cestari, 1982), and sophisticated control and measuring equipment (see Section 3.3).

2.1.4   Rods

The standard push rod is made of high tensile steel, and it is 36 mm outer diameter (usually 16 mm inner diameter), and of length 1 m (Figure 5). It is required to have a smooth finish with no protruding edges internally at joints. The diameter of the standard inner rod is specified as between 0.5 and 1.0 mm less than the internal diameter of the push rods, and it is usually 15 mm. The inner rods should slide easily through the push rods, and polishing is sometimes used to ensure this.

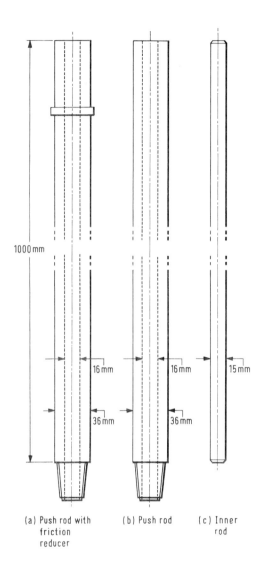

1000 mm

16 mm   16 mm   15 mm

36 mm   36 mm

(a) Push rod with   (b) Push rod   (c) Inner
    friction   rod
    reducer

Figure 5 Penetrometer rods

With light rigs, smaller diameter push rods are sometimes used (28 mm instead of 36 mm) but the push rods must have the same diameter as the cone over the 1-m length immediately behind the cone. In soft clays, aluminium inner rods are sometimes used.

With high penetration resistances, there is a danger of buckling of the push rods. It may therefore be necessary to support these within a casing where they are out of the ground or pass through soft clays.

2.2 Penetrometer tips

Penetrometers are of two main categories: mechanical and electrical. In mechanical penetrometers, the forces to mobilise cone resistance and local side friction are applied to the tip through rods and measured at the surface. With electric penetrometers, penetration is achieved by the application of force to the push rods. Forces are measured by electrical devices built into the tip, and the measurements are transmitted to the surface through a cable or by other means. There are also penetrometers in which forces are measured by hydraulic or pneumatic devices built into the tip. These are less common.

Mechanical and electric penetrometers can generally be further divided into those for measurement of cone resistance only and those for measurement of both cone resistance and local side friction. Another type, now little used, is that with which total side friction over the whole length of the push rods is measured by deducting the cone resistance from the total penetration resistance. This procedure is not recommended.

A section through an electric penetrometer tip, with a friction sleeve, is shown in Figure 6. Advantages of the electric penetrometer include:

1. accuracy and repeatability of results, particularly in weak soils
2. better delineation of thin strata

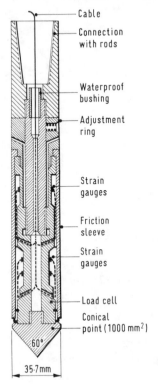

Cable

Connection
with rods

Waterproof
bushing

Adjustment
ring

Strain
gauges

Friction
sleeve

Strain
gauges

Load cell

Conical
point (1000 mm²)

60°

35·7mm

Figure 6 Section through an electric penetrometer tip (R)

3. faster over-all speed of operation
4. the possibility of incorporating pore pressure measurement or additional sensors in or above the tip (see Appendix B)
5. more manageable data handling.

Testing with electric penetrometers can be more expensive than with mechanical penetrometers, which are simpler and more rugged. A mechanical cone is sometimes used as a preliminary where the nature of the ground is uncertain.

Electric tips of different sensitivity are available, and it is important to select a tip appropriate to the maximum resistance likely to be met in a given profile. Where resistance varies widely, loss of accuracy occurs at lower resistances. A recent development which overcomes this problem is the 'Brecone' (Rigden *et al.*, 1982). This incorporates two load cells, of different load ranges, which simultaneously register resistance. Mechanical overload stops are provided to isolate each of the load cells once its maximum design working load has been reached.

2.2.1   Reference test tip geometry

The reference tip (R) (see Figure 1, page 5) has a common diameter of 35.7 mm for the base of the cone and the friction sleeve. The push rods should have this same diameter over the 1-m length above the base of the cone (Figure 1 of Appendix D). (Similar recommendations have been made by ASTM D3441 – 75T, 1975.)

Other standardised features for the reference tip include the height of the shoulder above the base of the cone and of the gap, containing a seal, above the shoulder of the cone (see Figure 1).

2.2.2   Tips not conforming to reference test geometry

A simple penetrometer tip is shown in Figure 8 of Appendix D. Because of the possibility of jamming of soil between cone and push rod, this type of mechanical penetrometer tip was largely superseded by the Dutch mantle cone tip (Figure 7(a)). It can be seen that the mantle is smaller in diameter than the base of the cone, and that it is slightly tapered.

The Dutch friction sleeve penetrometer tip which followed is shown in Figure 7(b). Here, although the mantle is parallel sided, both it and the friction sleeve are smaller in diameter than the base of the cone. The Delft electric penetrometer tip and the Delft friction sleeve electric penetrometer tip, which are shown in Figure 10 of Appendix D, also have geometries which differ considerably from the reference tip. These tips are now rarely used.

2.2.3   Effect of differing tip geometries

Both cone resistance and local side friction are influenced by the geometry of the penetrometer tip. Rol (1982) pointed out that experience shows that the cone resistance measured in sand with the mantle cone tip can be up to 10% higher than

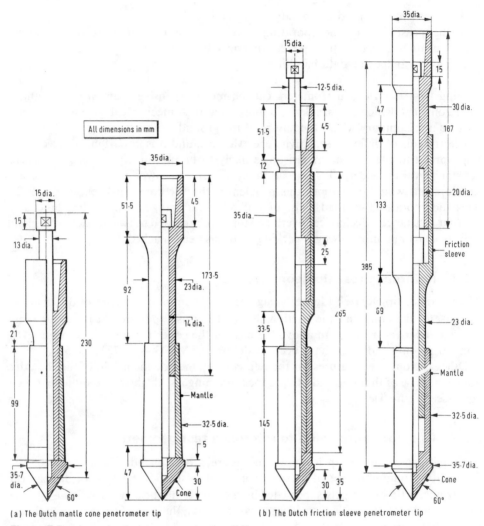

(a) The Dutch mantle cone penetrometer tip   (b) The Dutch friction sleeve penetrometer tip

Figure 7 Dutch mechanical penetrometer tips (M)

that measured with a simple cone. He found significant differences in comparisons of cone values in sand for three penetrometer tips:

1. an electric friction sleeve tip (R)
2. a Dutch mechanical friction sleeve tip (M)
3. a Delft electric friction sleeve tip (E1.2).

The designations R, M, E1.2, are those used by the ISSMFE sub-committee in their report (given in Appendix D). The tests showed that the ratio of E1.2 cone resistance to electric (R) cone resistance varied between 0.62 and 0.82, average 0.72*. The ratio appeared not to vary with cone resistance.

The ratio of the cone resistance of the Dutch mechanical friction sleeve tip (M), to the cone resistance of the electric friction sleeve tip (R) varied from 0.85 for a combination of low cone resistance values and shallow depth up to about 1.1 for a combination of higher cone resistance and greater depth. The relationship is given by:

$$(q_c)_M/(q_c)_R = 0.85 + 0.00035(D+2)\,(q_c)_R$$

where D = depth in metres

and $(q_c)_M$ and $(q_c)_R$ are cone resistances in MN/m^2 measured by the Dutch mechanical friction sleeve tip (M) and the electric friction sleeve tip (R), respectively.

This would indicate that results from the mechanical penetrometer are affected by friction between the push rods and the inner rods. Some manufacturers are now producing a highly polished surface on the inner rods and the inner surface of the push rods which they claim virtually eliminates this effect (Van den Berg, 1982). Resistance may also build up on the lower edge of the friction sleeve of the Dutch mechanical friction sleeve tip.

The considerable distance between the cone and the friction sleeve of the Dutch mechanical friction sleeve penetrometer tip makes for imprecision in measurement of friction ratio values.

The effect of tip geometry on cone resistance in clays in discussed in Sections 6.2 and 6.3.

2.2.4   Effect of tip diameter

Penetrometer tips with larger end areas are sometimes used to increase measurement sensitivity in weak soils. Experience with cylindrical electric penetrometer tips with projected cone areas between 500 and 1500 mm² shows that they produce cone resistances and local side frictions which do not differ significantly from those given by the 1000-mm² tip (De Ruiter, 1982). De Beer (1977) found that, in stiff clays, cone resistance decreases with increasing cone diameter. With a 50-mm dia. cone, q_c is 14% less than with a 36-mm dia. cone.

2.3   Overwater penetrometers

In water depths to about 25 m, penetration testing can be carried out from a platform through an outer casing which provides support to the push rods, or from a floating craft by supporting the thrust machine on a strong casing mounted on a concrete block or heavy steel plate set on the sea bed.

In deeper water, CPTs are carried out using electric penetrometers, either through a drillpipe or using a remotely-controlled jacking frame mounted on the sea bed. When

---

* This ratio relates to two sites (Delft and Diemen) tested by Rol. Tests at two other sites (Heijnen, 1973), where depths were shallower and resistances lower, gave an average ratio of 0.67.

working through a drillpipe, the penetrometer is held in place during a test by a latching device, and then the drillpipe is advanced, by rotation and flush, to the next test position, after which the penetrometer is again latched (Zuidberg and Windle, 1979, Van den Berg, 1982). The reaction is provided by the weight of the drillpipe and collars or by a down-hole packer. Seabed devices include 'Seacalf' (Zuidberg, 1975), 'Stingray' (Semple and Johnson, 1979), 'Dotipus' (Noorany and Giziensky, 1970), and the 'Hyson seabed penetrometers' (Van den Berg, 1982). See also Bruzzi (1983). These devices operate to water depths of 500m or more, and they can achieve penetrations below the sea bed similar to those achievable on land.

In water depths to 200m, testing can also be done from a manned diving bell (Vermeiden and Lubking, 1978) with a 700-kN ram.

3 Procedure

3.1 Extent of investigation

Generally, CPTs are used in conjunction with borings and possibly trial pits. The design of an investigation (i.e. the depth of exploration and the spacing between points of exploration) is not discussed in this Report. For guidance on this aspect reference should be made to Section 10 of BS 5930 (1981).

The circumstances under which it may be justified to use CPTs alone are indicated in Section 1.2. Typical Dutch practice in such cases (Te Kamp, 1977) is to put down one CPT or every two dwellings, at a housing site, and CPTs at 10- to 15-m centres on industrial projects. The depth of exploration is taken as 1.5 times the width of the foundation area, with the additional requirement that, for isolated piles, the exploration should extend at least four pile diameters below pile toe level.

3.2 Operation of equipment

Some indication of operational procedure has been given earlier, but for ease of reference it is amplified below. This Section should be read in conjunction with Appendices C and D.

With the light penetrometer rig and a mechanical tip, the cone is advanced by hand operation of a chain drive at a rate as close to 20 mm/s as practicable. The load is measured at the top of the rods by a hydraulic load cell with two pressure gauges or with a proving ring and dial gauge. When a cone penetration has been completed by advancing the cone attached to the inner rods, the push rods are advanced until they rejoin the cone and the whole assembly is pushed down to the next cone test level. Tests are commonly carried out at 200-mm intervals. With a light rig, total side friction is sometimes evaluated by measuring the force necessary to advance the push rods and cone together and deducting from this the force necessary to advance the cone alone. On the other hand, if no measurement of total skin friction is to be made (as is generally the case), a friction reducer and/or push rods of a smaller diameter are used. Local skin friction is sometimes measured with a light penetrometer rig, in the way described below for a medium rig.

With a medium penetrometer rig, penetration is obtained by use of a hydraulic ram or rams. Usually, local side friction is measured in addition to cone resistance. The standard rate of testing is 20 ± 5 mm/s. A mechanical tip is used in most cases, although electric tips are sometimes used with 100-kN rigs.

Where the electric penetrometer tip is used (as is generally the case with heavy rigs), it is also advanced at a rate of 20 mm/s, but a continuous reading is made of cone resistance, and of local side friction where a friction sleeve is incorporated. The cable connecting the load cells within the tip to the read-out instruments is pre-threaded through the push rods, sufficient rods being threaded to achieve the expected depth of testing.

With a purpose-built truck rig and a heavy penetrometer, the rate of testing can be between 150 and 300 m per day. With a trailer-mounted medium rig, requiring screw anchors to provide reaction, the rate of penetration is likely to be less than one third of this rate. A picketed light rig can probably achieve a maximum of about 50 m per day.

3.3 Recording results

The various possibilities for recording and processing CPT data are shown schematically in Figure 8.

With most mechanical penetrometers, readings taken from the hydraulic pressure gauge or from the proving-ring dial gauge are recorded manually (Path A in Figure 8). A typical field record sheet is shown in Figure 9. Subsequent data processing and

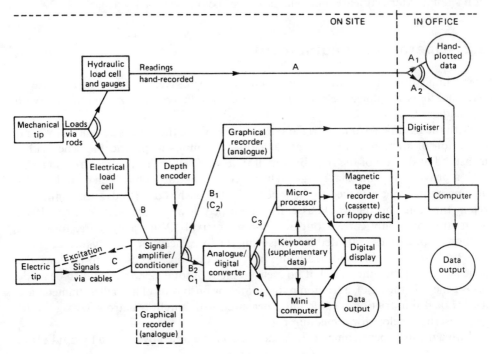

Figure 8 Possible arrangements for recording and processing CPT data

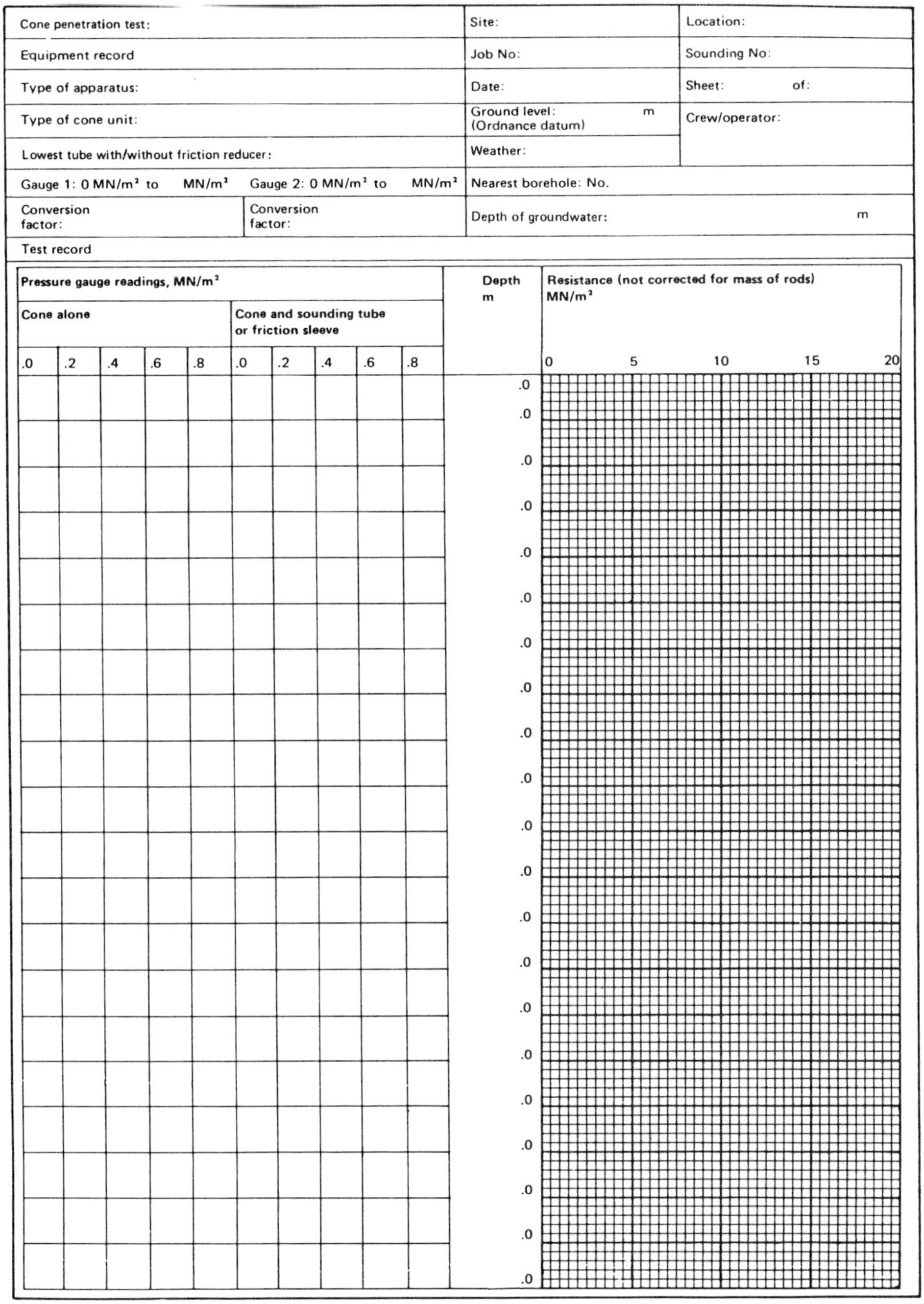

Cone penetration test:		Site:	Location:
Equipment record		Job No:	Sounding No:
Type of apparatus:		Date:	Sheet: of:
Type of cone unit:		Ground level: m (Ordnance datum)	Crew/operator:
Lowest tube with/without friction reducer:		Weather:	
Gauge 1: 0 MN/m² to MN/m² Gauge 2: 0 MN/m² to MN/m²		Nearest borehole: No.	
Conversion factor:	Conversion factor:	Depth of groundwater: m	
Test record			

Pressure gauge readings, MN/m²										Depth m	Resistance (not corrected for mass of rods) MN/m²
Cone alone					Cone and sounding tube or friction sleeve						
.0	.2	.4	.6	.8	.0	.2	.4	.6	.8		0 5 10 15 20

Figure 9 Example of field record sheet, using a mechanical penetrometer tip (after BS 5930: 1981)

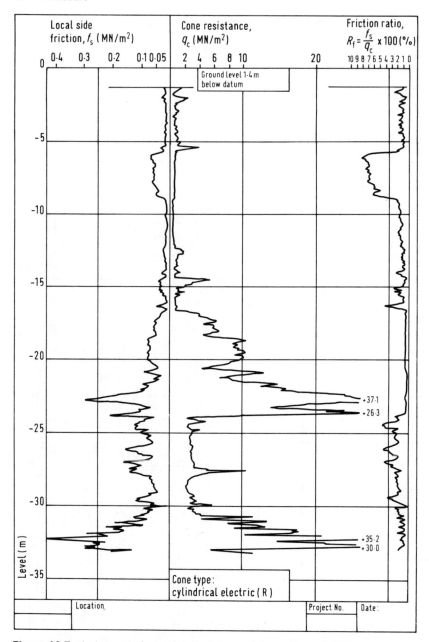

Figure 10 Typical record of cone penetration test (after te Kamp, 1977)

plotting can be done manually (Path A1) or by computer (Path A2). In addition to the plots against depth of cone resistance, q_c, local side friction, f_s, and the ratio (f_s/q_c) between these (the 'friction ratio'), the record sheet should contain such essential information as the location and date of the test, and ground level at the test position. An example is given in Figure 10.

However, with some mechanical penetrometers the loads transmitted by the rods are measured electrically and fed into a signal amplifier/conditioner unit (Path B). They are then either plotted on an analogue chart recorder for subsequent digitising and computer processing in the office (Path B1), or treated in the same way as signals from an electric penetrometer tip (Path B2) as outlined below.

With an electric penetrometer tip, the signals are fed into a signal amplifier/conditioner, and monitored by means of an analogue graphical recorder. The conditioner unit provides the excitation voltages to the penetrometer tip. It also receives pulses from a depth encoder so that depth can be accurately measured. As with electrical signals from a mechanical tip, the records from the graphical recorder can be digitised for computer processing (Path C2), but generally the conditioned signals are passed to an analogue/digital (A/D) converter (Path C1) then to a microprocessor (Path C3), where they are put on to magnetic tape for computer processing when the cassettes are returned to the office. The micro-processor is usually provided with a digital display unit and a keyboard, so that supplementary data such as the site and test location can be added to the tape. An alternative is to use a mini-computer on site (Path C4) to process the data fully. Whether the computer is site or office based, output can be in a variety of forms, as required.

As an alternative to cable connection between the tip and the surface, acoustic signals can be used, or the tip can be fitted with a solid-state memory (Muromachi, 1981). This latter case has the disadvantage that the test cannot be monitored as it proceeds.

3.4 Accuracy and calibration

Cone resistance varies with the rate of penetration. In sands, the variation involved in the generally accepted $\pm 5\,\text{mm/s}$ tolerance on the standard rate of $20\,\text{mm/s}$ is negligible. There is a special situation with very loose, saturated sands, which may liquefy (or almost liquefy) when penetrated at the standard rate, giving very low cone resistance. The effect of variation of rate of penetration in clays is discussed in Section 6.1.

Measurements become less accurate if the dimensions of the cone depart appreciably from the standard dimensions through wear or by damage. Frequent checks should therefore be made to see that divergences do not exceed the tolerances set out in Appendix D. Of particular importance is the surface roughness of the cone. A newly-manufactured cone has a surface friction angle, δ, of about 0.5ϕ, where ϕ is the angle of shearing resistance of the soil. A 50% increase, to 0.75ϕ gives rise to an increase in penetration resistance of about 50% (Dorgunoglu and Mitchell, 1975). The sleeve friction is affected by roughness of the sleeve. The recommendation for a standard (Appendix D) requires that the sleeve shall have, and maintain with use, a roughness of $0.5\mu\text{m} \pm 50\%$ (Appendix D, 3.5 and 11).

The cone (and friction sleeve where used) should be checked for free sliding both at the start and at the end of each penetration. All push rods and inner rods should be straight and clean, and push rods should be well oiled internally. Before each test, the seals between different elements of a tip should be cleaned and inspected to check their integrity.

Load cells used with mechanical tips should be calibrated at the start and finish of a programme, and at regular intervals during a programme where this involves more than a few soundings.

Electric tips provide more accurate and repeatable results than mechanical tips, but they are subject to zero-load errors and calibration errors, both of which tend to change during testing. The zero-load error can readily be checked by observing the zero-load deflection before and after each test. It should not exceed 1 to 2% of the rated load (De Ruiter, 1982).

The load measurement system should be calibrated at intervals not exceeding 3 months (more frequently when used continuously), also after every overhaul or repair. Calibration should be against a proving device with accuracy confirmed against a National Standard. The calibration should be based on at least ten load increments within the working load range of the penetrometer tip. It should be carried out at 25°C, and checked over the working temperature range of the tip. Calibration should be checked daily in the field with an appropriate field control unit.

For most fieldwork applications, the aim should be for errors in cone resistance and side friction readings not to exceed about 3%. In an electric tip, it is usual to provide compensation for temperature variations.

Loss of accuracy can occur when testing soft clays with an electric tip, because readings extend only over the lower part of the scale range. It is therefore preferable to have a separate tip for soft clays if their resistance is to be accurately measured, or to have a tip with two measuring ranges (see Section 2.2).

With mechanical penetrometers, the weight of the inner rods has to be allowed for when testing weak soils, and in soft clays the rods may be too heavy. Aluminium rods help in such a situation.

The depth to the penetrometer tip should be measured within 25mm. Inaccuracy in measurement of friction ratio can occur because of the difference in level between cone and friction sleeve, particularly with the Dutch mechanical friction sleeve tip and the Delft electric friction sleeve tip. With a mechanical penetrometer, there is difficulty in matching cone resistance to side friction at the same level because of irregularity of the readings. With an electric tip, where a depth encoder is used, reasonably accurate combination can be made of cone resistance and side friction at mid height of the friction sleeve.

A further source of error arises if penetration deviates appreciably from the vertical. This can be checked with an electric tip which includes a slope sensor, which should be accurate to ± 1%.

To avoid increased friction between push rods and inner rods, a mechanical cone should not be used for depths greater than about 20m.

To avoid disturbed ground, a CPT should not be performed within a distance from a borehole less than 25 times the borehole diameter, nor within 1m of a previous sounding.

4 Use of CPT results

4.1 General

During a cone penetration test, complex changes take place in stresses, strains and pore pressures, making comprehensive theoretical analysis difficult. Hence, although theoretical analyses of the CPT are available, in practice the use of the CPT remains essentially empirical.

The role of the CPT was outlined in Section 1.2. Its use in profiling and identification of soils is discussed in Section 4.2, together with the basis for assessing engineering parameters. The main relevant parameters are: angle of shearing resistance (Section 5.3) and deformation characteristics (Section 5.4) in cohesionless soils, and undrained shear strength (Section 6.3) and modulus (Section 6.4) in cohesive soils. Parameters in other materials are discussed in Section 7. Practical applications are covered in Sections 8, 9 and 10. These include assessment of the ultimate bearing capacity of piles (Section 9.1), which can be made by direct empirical use of CPT results without intermediate determination of engineering parameters.

In theory, it is necessary to correct cone resistance and local side friction for pore pressure effects, but in practice such corrections have not been used in published data. This is discussed in Section 12.1.1.

Caution is required when using published correlations between CPT results and engineering parameters, and care is needed to check that correlation is obtained from a tip of similar geometry (e.g. correlations are sometimes presented as applying to sands in general, whereas their applicability may be limited to the class of sand tested: angularity, particle size and grading, mineral content, over-consolidation ratio, etc.). Similarly, correlations based on London Clay, about which much is known, may not be applicable to other stiff, fissured clays. Wherever possible, site-specific correlations should be developed.

In this Report, more than one approach is described for the evaluation of an engineering parameter, or more than one calculation method is presented. Where no preference is indicated, it is suggested that all the approaches should be used, and that judgement should then be exercised in deciding the value to be adopted. When CPT data are used to derive parameters for a geotechnical design, the design should generally be checked by an alternative approach (e.g. other *in-situ* tests, laboratory tests, or field loading tests). The circumstances under which it may be justified to use CPTs alone are outlined in Section 1.2.

4.2   Profiling and identification of soils

One of the important uses of the CPT is to delineate the soil profile. This it can do with greater accuracy than can be achieved from conventional boring and sampling. However, the accuracy is not greater than can be obtained in certain soils with a sampler such as the Delft Continuous Sampler (see Appendix B, 2.1), or by continuous sampling with a stationary piston sampler. Moreover, such sampling has the advantage that it permits visual inspection of the soil.

Cone resistance responds to soil changes within 5 to 10 tip diameters above and below the cone, the distance increasing with increasing soil stiffness. This leads to some imprecision in locating soil interfaces, and it has some effect on the evaluation of engineering parameters (e.g. cone resistance may not reach its full value in a sand layer bounded by soft clay layers if the sand layer is less than about 0.7 m thick). Very thin layers can be missed. A thin layer of sand within a clay stratum may not be detected if it is less than about 100 mm thick, and a clay layer within sand may not be detected if it is less than 150 to 200 mm thick. Greater accuracy can be obtained by using a pore-pressure probe or a penetrometer tip fitted with a piezometer (i.e. a piezocone) (see Section 11.5).

A broad identification of soils can be obtained from the magnitude of the cone resistance, and more specifically from their friction ratios (i.e. the ratio of local side

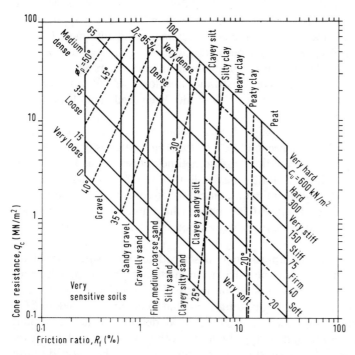

Figure 11 Identification of soils using the Dutch mechanical friction sleeve penetrometer (from Searle, 1979)

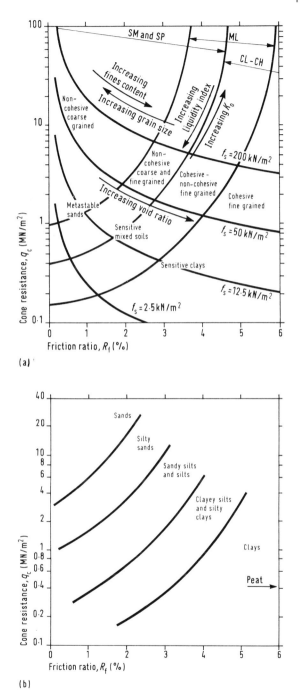

Figure 12 Identification of soils using the reference test penetrometer tip (R). (a) Full scheme (after Douglas and Olson, 1981). (b) Working version (after Robertson and Campanella, 1983)

friction to cone resistance) at the same level, considered in relation to the cone resistance. A scheme for identification using the Dutch mechanical friction sleeve tip was first formulated by Begemann (1965) and extended by Schmertmann (1969). Searle (1979) incorporated the results of further field measurements and produced the version shown in Figure 11*. A comprehensive scheme for use with cones conforming with the reference test tip geometry (R) by Douglas and Olsen (1981) is shown in Figure 12(a), and a simplified working version by Robertson and Campanella (1983) is given in Figure 12(b). There is some evidence to show that the friction ratio for some fine-grained soils may decrease with increasing effective overburden pressure. Thus the accuracy of Figures 11 and 12 may be reduced with very deep soundings (Robertson and Campanella, 1983).

It is advisable to check soil identification by the CPT by direct comparison against one or more boreholes, preferably with continuous sampling. It should also be emphasised that local experience may well indicate differences from the data presented above. Every opportunity should be taken to derive correlations based on local conditions.

---

* For fresh-water alluvial and lacustrine silts and clays in Northern Italy, Cancelli (1983) found somewhat lower R_f values than those given in Figure 11. The difference was greater in overconsolidated soils than in normally-consolidated soils.

5 Parameters in cohesionless soils

5.1 General

The CPT has an important role in the exploration of cohesionless soils, because there is a lack of satisfactory alternative methods. Laboratory testing is generally not feasible, because of the difficulty of obtaining undisturbed samples. Alternative *in-situ* tests include the plate-loading test and the screw-plate loading test (field compressometer) (Janbu and Senneset, 1973). Both tests are time consuming, and plate-loading tests are difficult and expensive below the groundwater level.

Direct derivations from cone resistance of relative density, angle of shearing resistance, and modulus values (either the constrained modulus or Young's modulus) depend on empirical correlations. These have some backing from calibration chamber tests, but it should always be remembered that such correlations are limited in the range of soils to which they apply.

CPT results can also be transposed into equivalent SPT N values in order to use correlations between N and engineering parameters and performance. This is generally less satisfactory than direct correlation.

5.2 Relative density

Relative density is difficult to measure in laboratory tests, because of the uncertainties involved in maximum and minimum density determinations (ASTM, 1973). Nevertheless, relative density is sometimes used as an intermediate parameter or as a general guide in design. Most knowledge of the relationship between relative density, D_r, and cone resistance, q_c, comes from calibration chamber tests, which usually require correction for the effects of chamber size.

5.2.1 Normally-consolidated (NC) sands

The relationship between relative density, D_r, and cone resistance, q_c, of a sand is greatly affected by its compressibility. For a given value of relative density and effective overburden pressure, σ'_{vo}, a sand of high compressibility has a lower q_c than a sand of low compressibility. Figure 13 shows values of D_r plotted against a function

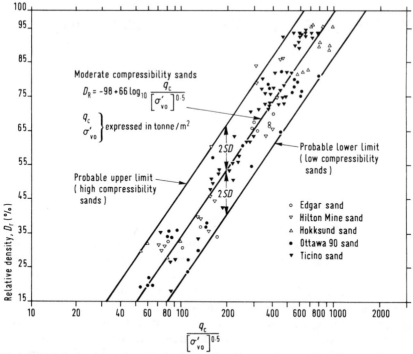

Figure 13 Relationship between relative density and cone resistance of uncemented, normally-consolidated quartz sands (after Jamiolkowski et al., 1985) (Note: Correction required for chamber size)

of q_c and σ'_{vo}. These values derive from calibration chamber tests (Jamiolkowski et al., 1985), and they require correction for chamber size, as detailed below. Some properties of the sands used in the tests are given in Table 1.

Figure 13 shows that, whereas the sands of moderate compressibility (Ticino, Edgar, Hokksund) follow the regression line (albeit with a good deal of scatter), the results

Table 1   Some properties of the sands in Figure 13

Sand	Mineralogy	Shape	Grading (mm)		Porosity	
			D_{60}	D_{10}	n_{max}	n_{min}
Ticino	Mainly quartz, 5%* mica	Sub-angular to angular	0.65	0.40	0.50	0.41
Ottawa 90	Quartz	Rounded	0.24	0.13	0.44	0.33
Edgar	Mainly quartz	Sub-angular	0.50	0.29	0.48	0.35
Hokksund	35% quartz 45% feldspar 10%* mica	Rounded to sub-angular	0.50	0.27	0.48	0.36
Hilton Mine	Quartz + mica + feldspar	Angular	0.30	0.15	0.44	0.30

*Percentage mica by volume

for the highly compressible Hilton Mine sand tend towards the upper line, and those for the Ottawa 90 sand, of low compressibility, tend towards the lower line. It is suggested, therefore, that these upper and lower lines on Figure 13 can be taken to represent probable limit values of relative density for sands of very high compressibility and very low compressibility, respectively.

The correction for chamber size is made by dividing q_c by K_q, where the correction factor, K_q, is given by:

$$K_q = 1 + \frac{0.2(D_r - 30)}{60} \qquad\qquad (1)$$

Because K_q is a function of D_r, it is necessary to determine D_r from Figure 13 and Equation (1) on a trial and error basis.

Figure 13 applies to relatively uniform, uncemented, clean, NC, predominantly quartz sands, where the *in-situ* horizontal stress ratio, K_o, is about 0.45. In using Figure 13, the probable compressibility of the sand should be taken into account, bearing in mind that compressibility is greater where the sand is uniform in grading, where the sand grains are angular, and where there is an appreciable mica content.

In a thin sand layer, an under estimate of D_r may be obtained, because the full cone resistance may not have developed.

5.2.2 Overconsolidated (OC) sands

Chamber tests show that, for a given sand, D_r is better correlated with initial horizontal effective stress, σ'_{ho} than with initial vertical effective stress, σ'_{vo}. Figure 13 can therefore be used with OC sands, if the *in-situ* effective horizontal stress is known or can be estimated, by substitution of σ'_{ho} for σ'_{vo}. This should be regarded as only an approximation, with a possible error in D_r of $\pm 20\%$. As for NC sands, a correction for chamber size is required. The presence of OC sands can sometimes be indicated if relative densities greater than 100% are derived using the procedures set out in Section 5.2.1.

Values of *in-situ* horizontal stress can be estimated in the field using a self-boring pressuremeter, or, more conveniently, on a CPT site, using a flat dilatometer (see Appendix B). Failing this, and provided an estimate of the overconsolidation ratio (OCR) is available, values of K_o (and hence of σ'_{ho}) can be estimated using Schmertmann's relationship (1975):

$$K_{o(OC)} = (OCR)^{0.42}\, K_{o(NC)}$$

Schmertmann (1975) also suggests a relationship between q_c for OC sand and q_c for NC sand of the form:

$$q_{c(OC)}/q_{c(NC)} = 1 + \chi[(OCR)^\beta - 1]$$

where χ and β have the values 0.75 and 0.42, respectively. Baldi *et al.* (1983) found that β varies with relative density:

$$\beta \simeq 0.275 + 0.26 D_r,$$

and that χ varies approximately from 0.50 (OCR = 2) to 0.25 (OCR = 15), decreasing with increasing D_r.

5.3   Strength

It is possible to estimate the peak effective angle of shearing resistance, ϕ', of free-draining sands using relative density as an intermediate parameter, taking values from Figure 13, and the relationships between ϕ' and D_r given by Schmertmann (1978) as shown in Figure 14. Another approach is to use the Terzaghi bearing capacity factor for general shear, N_γ, as an intermediate parameter. A correlation between N_γ and q_c is given by Muhs and Weiss (1971):

$$N_\gamma = 12.5q_c \quad (q_c \text{ in MN/m}^2)$$

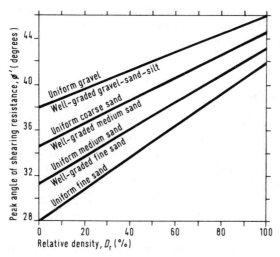

Figure 14 Relationship between peak angle of shearing resistance and relative density of quartz sands (after Schmertmann, 1978)

This correlation was derived from large-scale shallow footing tests on sand, and it takes no account of overburden pressure. These procedures are less satisfactory than direct procedures for determination of ϕ', because of the double uncertainty involved.

A direct correlation between q_c and ϕ' is shown in Figure 15. It is derived from a bearing-capacity theory developed by Durgunoglu and Mitchell (1975), using a soil-to-cone friction angle equal to $0.5\phi'$ and a lateral earth pressure coefficient, $K_o = 1 - \sin \phi'$. The theory ignores the effects of soil compressibility.

A review by Robertson and Campanella (1983) of the results of chamber correlation tests by a number of researchers showed that the correlation provided by the Dorgunoglu and Mitchell theory gives a reasonable lower bound of ϕ' for sands of the type tested. These were predominantly of quartz, with some feldspar and in some

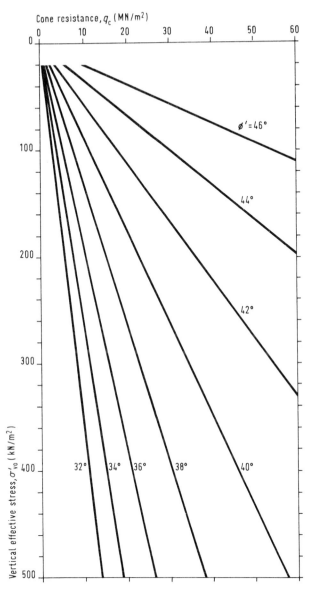

Figure 15 Relationship between angle of shearing resistance and cone resistance for an uncemented, normally-consolidated quartz sand (after Durgunoglu and Mitchell, 1975)

cases a small amount of mica. Particle shapes ranged from rounded to sub-angular. The sands were in the medium and medium-to-coarse particle size range and fairly uniform in grading (uniformity coefficients from 1.6 to 2.2). In practice, clean NC sands of this kind have ϕ' ranging from the lower-bound values given in Figure 15 up to about 2° higher for the more compressible ones (angular grains, higher content of mica, more uniform).

In highly compressible sands (e.g. carbonate sands or glauconitic sands), ϕ' may be significantly higher than would be derived from Figure 15. However, the presence of compressible sands can be detected from their friction ratios. If R_f (from the reference tip) exceeds about 0.5%, Figure 15 probably underestimates ϕ'. (Some carbonate sands have R_f values as high as 3% (Joustra and de Gijt, 1982), but this may not apply in cemented carbonate sands, where a reduction in R_f would be expected.) The ϕ' of cemented sands also is underestimated by Figure 15.

The method of Senneset and Janbu (1985) (set out in Section 12.2.3) requires measurement of pore pressure during the CPT. However, it can be used for estimating the angle of shearing resistance of free-draining sands without the need for pore-pressure measurements.

5.3.1 Effect of overconsolidation

The use of Figure 15 for OC sands overestimates the secant angle, ϕ', by 1 or 2°.

5.3.2 Curvature of the strength envelope

The theory underlying angles of shearing resistance presented in Figure 15 takes no account of the curvature of the strength envelope. It should be borne in mind that, at higher confining stresses, ϕ' is somewhat lower, the difference increasing with increasing relative density. Very approximately, a one-log cycle increase in confining produces a decrease in ϕ' as follows:

$$D_r < 0.35 \qquad 0° \text{ to } 1°$$
$$0.35 < D_r < 0.65 \qquad 2° \text{ to } 3°$$
$$0.65 < D_r < 0.85 \qquad 3° \text{ to } 5°$$
$$0.85 < D_r \qquad 5° \text{ to } 8°$$

Somewhat larger reductions may occur in sands of high compressibility.

5.4 Deformability

Depending on the problem under consideration, it may be necessary to evaluate one of three moduli: the constrained modulus, M (which is equal to the reciprocal of the oedometer vertical coefficient of volume change, m_v), the Young's modulus, E, or the shear modulus, G. Furthermore, because stress-strain curves for sands are non-linear, it is necessary to fix a stress range over which the modulus is to be determined. Diagrams from which M, E and G of NC, uncemented, predominantly quartz sands can be estimated as a function of q_c and σ'_{vo} are given in Figures 16, 17, and 18, respectively. These derive from chamber tests. They may significantly underestimate the modulus values of OC sands.

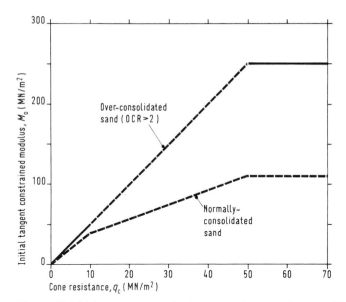

Figure 16 Initial tangent constrained modulus for normally-consolidated sands (after Lunne and Christoffersen, 1983)

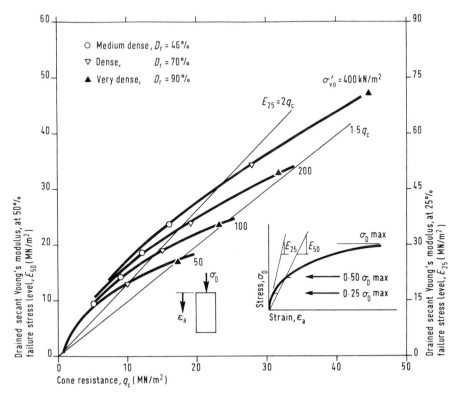

Figure 17 Secant Young's modulus values for uncemented, normally-consolidated quartz sands (after Robertson and Campanella, 1983, based on data from Baldi *et al.*, 1981)

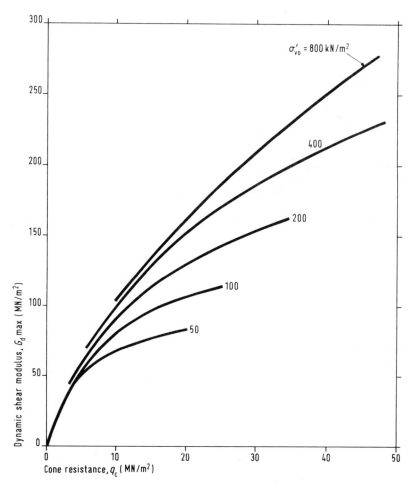

Figure 18 Dynamic shear modulus for uncemented, normally-consolidated, predominantly quartz sands – small strains (after Robertson and Campanella, 1983)

5.4.1 Constrained modulus

Correlations between constrained modulus, M, and cone resistance, q_c, are commonly expressed in the form:

$$M = \alpha_M \cdot q_c$$

and α_M is often stated to be in the range 1.5 to 4. Vesic (1970) suggests:

$$\alpha_M = 2 \left(1 + \left(\frac{D_r}{100} \right)^2 \right)$$

which gives α_M values in the range from about 2.25 to 4. Table 2 shows (Lunne and Kleven, 1981) α_M from a number of chamber tests. It can be seen that for NC sands, α_M lies in the range 3 to 11, with higher values for OC sands.

Table 2   Summary of calibration chamber results for constrained modular coefficient

Source	NC sand		OC sand	
	No. of sands	α_M	No. of sands	α_M
Veismanis (1974)	3	3 to 11	3	5 to 30
Parkin et al. (1980)	1	3 to 11	1	5 to 30
Chapman & Donald (1981)	1	3 to 4	1	8 to 15
		3 absolute lower limit		(12 = average)
Baldi et al. (1982)	1	> 3	1	3 to 9

In practice, α_M decreases with increasing q_c, but for a given value of q_c, α_M increases with increasing stress level.

Webb et al. (1982) suggest that M values should be calculated as follows:

Clean sands $\qquad M = 2.5(q_c + 3.2) \text{ MN/m}^2$

Clayey sands $\qquad M = 1.7(q_c + 1.6) \text{ MN/m}^2$
(clay content
   about 20%)

This is based on some earlier work (Webb, 1969), where settlements were measured under large plate-loading tests, and confirmed in the later work where settlements were measured over intervals of depth below a fill of large lateral extent. Cone resistances were generally low (2 to 5 MN/m²), and within this range the corresponding α_M values are between 4 and 6.5 for clean sand and between 2 and 3 for clayey sand.

Lunne and Christoffersen (1983) made an extensive review of chamber test results, of field tests carried out at the Norwegian Geotechnical Institute, and of Webb's field tests. They propose conservative values of the initial tangent constrained modulus, M_o, in NC sand:

$M_o = 4q_c$ $\qquad$ for $\qquad\qquad\qquad q_c < 10 \text{MN/m}^2$
$M_o = 2q_c + 20$ $\qquad$ for $\qquad 10 \text{MN/m}^2 < q_c < 50 \text{MN/m}^2$
and
$M_o = 120 \text{ MN/m}^2$ $\quad$ for $\qquad 50 \text{ MN/m}^2 < q_c$

For OC sand, with OCR > 2, they propose:

$M_o = 5q_c$ $\qquad$ for $\quad q_c < 50 \text{MN/m}^2$
and
$M_o = 250 \text{ MN/m}^2$ $\quad$ for $\quad q_c > 50 \text{MN/m}^2$

These recommendations are shown in Figure 16.

The constrained modulus applicable for the stress range σ'_{vo} to $\sigma'_{vo} + \Delta\sigma_v$ can be estimated as:

$$M = M_o \left(\frac{\sigma'_{vo} + \Delta\sigma_v/2}{\sigma'_{vo}} \right)^{0.5} \qquad\qquad (2)$$

This relationship is applicable to NC sands, but for OC sands the exponent decreases with OCR approaching zero for heavily OC sands.

5.4.2 Young's modulus

For other than one-dimensional cases, Young's modulus, E, is more appropriate than the constrained modulus, M. As with M, E is dependent on stress level, and for NC sands Figure 17 gives values of drained secant modulus at 25% of failure stress (E_{25}) and at 50% of failure stress (E_{50}), based on chamber test results. (It can be seen that these are at a constant ratio.) For most foundation problems, E_{25} is relevant, although E_{50} may be more relevant when considering end-bearing capacity of piles. Figure 17 shows that, except at very low relative densities, E_{25} varies between about $1.5q_c$ and just over $2q_c$, which is in reasonable agreement with the value of 2.5 recommended by Schmertmann (1970) for calculation of settlements of footings on sand.

For OC sands, E_{50} varies between $6q_c$ and $11q_c$ (Baldi *et al.*, 1982). E_{25} is some 50% higher for OCRs up to about 3, but approximately equal to E_{50} for OCR greater than 4 (ENEL *et al.*, 1985). However, as a result of the limited data available, for OC sands, it is prudent to adopt E values not more than twice those given in Figure 17 for NC sands.

5.4.3 Dynamic shear modulus

Figure 18 shows correlations obtained by Robertson and Campanella (1983) between dynamic shear modulus, cone resistance, and vertical effective stress. It is based on laboratory correlations between dynamic shear modulus, at small strains (less than 10^{-3}% dynamic strain amplitude), and relative density (Seed and Idriss, 1970, Hardin and Drnevich, 1972), and the relationship between D_r and q_c developed by Baldi *et al.* (1981).

Dynamic shear modulus at any strain level can be estimated using the Seed and Idriss reduction curves (Seed and Idriss, 1970). The initial tangent static modulus is about one fifth of the dynamic modulus at small strains (Byrne and Eldridge, 1982).

5.5 CPT and SPT

There is a considerable body of knowledge relating the calculation of bearing capacity and settlement of foundations on cohesionless soils to SPT N values (Nixon, 1982). One route therefore in such calculations is to transpose cone resistance values, q_c, into SPT N values. Meyerhof (1956) proposed a relationship for fine sand which in SI units becomes:

$$q_c = 0.4N$$

where q_c is in MN/m^2 and N is the number of blows for 0.3 m penetration.

Meigh and Nixon (1961) showed that the ratio between q_c and N varied from about 0.25 for silty fine sands up to 1.2 or more for coarse gravels. Various relationships have since been proposed by others (Thorburn, 1970, Schmertmann, 1970).

Burbidge (1982) collected together available data concerning the relationship between q_c/N and average grain size, D_{50}, including the data associated with case records referred to by Burland and Burbidge (1985). The resulting average correlation is presented in Figure 19, which also shows the zone within which the individual results fall.

After a comprehensive review, Robertson *et al.* (1982) related q_c/N to D_{50}, in the range $0.002 < D_{50} < 1\,\text{mm}$. Over this range, the Robertson and Campanella correlation is almost identical to Burbidge's.

The ratio of q_c to N in Figure 19 shows considerable scatter. Because of this uncertainty, direct use of CPT results is preferable to conversion to N values.

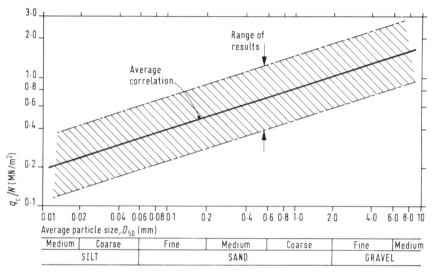

Figure 19 Relationship between cone penetration test and standard penetration test (from Burland and Burbidge, 1985)

A difficulty arises in converting cone resistances to N values in that N varies with the method of carrying out the test. In theory, N should be measured with a free-falling hammer. In US practice, N is measured either by releasing a slip rope which is taken two turns round a 'cat-head' (sometimes only one turn). In the UK, practice is to use a sliding hammer released by a tripping device, a 'trip hammer'. Field trials by Frydman (1970), using a cat-head with two turns, gave N values about 1.4 times those obtained with a trip hammer. Somewhat similar results were obtained by Serota and Lowther (1973) and Komornik (1974). However, recent work described by Baldi *et al.* (1985) suggests that the difference between N as measured by trip hammer and N measured by slip rope is not appreciable. A further complication arises in that two different hammers are in use in the USA, a 'safety' hammer and a 'donut' hammer. A safety hammer delivers to the rods about 60% of the free-fall energy, which is similar to that delivered by trip hammers in use in the UK. On the other hand, the 'donut' hammer delivers only about 45% of the free-fall energy (Seed *et al.*, 1985). The fact that there may be a difference is not very important when converting from q_c to N for foundation calculations, because Figure 19 is mainly based on slip-rope N values. Most of the empirical correlations between N and bearing capacity and settlement (see Section 8) are also based on the slip-rope method. However, should it be required to convert from N to q_c with N measured using a trip hammer, it should be appreciated that the resulting value of q_c may be underestimated.

In the preceding discussion, none of the N values mentioned in the correlations are corrected in any way.

5.6 Synopsis: parameters in cohesionless soils

5.6.1 Relative density

Refer to Section 5.2

Normally-consolidated sands: use Figure 13 making correction for chamber size by Equation (1), page 29

Figure 13 (Section 5.2.1) refers to relatively uniform, clean, uncemented, predominantly quartz sands.

- for sands of average compressibility, use the regression (central) line
- for sands of high and low compressibility, tend towards the upper and lower lines respectively.

Compressibility is greater where the sand is uniform in grading, where the sand grains are angular, and where there is an appreciable mica content.

In thin layers, the cone resistance, q_c, may lead to an underestimate of relative density, D_r.

Overconsolidated sands: as a first approximation enter Figure 13 with the horizontal effective stress, σ'_{ho}, derived from self-boring pressuremeter or flat dilatometer tests, or from

Refer to Section 5.2.2.

$$K_{o(OC)} = (OCR)^{0.42} \, K_{o(NC)}$$

- alternatively, use the relationship

$$\frac{q_{c(OC)}}{q_{c(NC)}} = 1 + \chi \, [(OCR)^{\beta} - 1]$$

where χ varies approximately from 0.50
   $(OCR = 2.0)$ to 0.25 $(OCR = 15)$
and β $\simeq 0.275 + 0.26 D_r$

5.6.2 Strength

Refer to Section 5.3.

Strength can be estimated indirectly from D_r (Figure 13) using Figure 14 for ϕ'

Sands are assumed to be free draining.

or from $N_\gamma = 12.5 q_c$ (q_c in MN/m^2)

Direct methods are preferred to indirect methods.

A direct estimation of strength may be made using Figure 15, which gives reasonable lower bound values of ϕ' for fairly uniform, moderately compressible, normally-consolidated, predominantly quartz sands. For overconsolidated sands, ϕ' is 1 or 2 degrees lower than would be estimated from Figure 15.

For more compressible sands, ϕ' is up to 2° higher.

For very compressible sands, ϕ' may be very much higher.

ϕ' decreases with increasing confining stress. Very approximately a one-log cycle increase in confining stress produces a decrease in ϕ' as follows:

$D_r < 0.35$	0 to 1°
$0.35 < D_r < 0.65$	2 to 3°
$0.65 < D_r < 0.85$	3 to 5°
$0.85 < D_r$	5 to 8°.

5.6.3   Deformability

Refer to Section 5.4.

(a) Constrained Modulus

Refer to Section 5.4.1.

For normally-consolidated sands

$$M = M_o \left(\frac{\sigma'_{vo} + \Delta\sigma_v/2}{\sigma'_{vo}} \right)^{0.5}$$

where M_o is the initial tangent modulus given by Figure 16.

For overconsolidated sands, the exponent reduces with OCR, approaching zero for heavily overconsolidated sands.

(b) Young's Modulus

Refer to Section 5.4.2.

For uncemented, normally-consolidated, quartz sands, E_{25} and E_{50} may be estimated from q_c, using Figure 17.

For overconsolidated sands, double the values from Figure 17.

(c) Shear Modulus

Refer to Section 5.4.3.

Dynamic shear modulus at small strains may be estimated from q_c, using Figure 18.

Initial tangent static modulus is about 1/5th of the dynamic modulus at small strains.

For dynamic shear modulus at any strain level use reduction curves, Seed and Idriss (1970)

5.6.4   CPT/SPT Conversion

Refer to Section 5.5 and Figure 19.

6 Parameters in cohesive soils

6.1 General

In contrast to cohesionless soils, cohesive soils form a less important field of use for the CPT, because established alternative methods are available. In NC and lightly OC clays, parameters can be obtained from *in-situ* vane tests and laboratory tests on specimens from thin-walled stationary piston samplers. Nevertheless, the CPT has the advantages of rapid coverage and good identification of variations in stratification. Clearly, a combination of CPT and alternative methods may be beneficial.

In OC clays, particularly stiff, fissured clays, vane tests are unsuitable, and laboratory tests for strength and elastic modulus suffer from the effects of sample disturbance. Unfortunately, the macrofabric of such clays also makes for difficulty in the interpretation of CPT results. For determination of constrained modulus, laboratory tests are preferable.

The cone resistance in clays varies with the rate of penetration. The available data in Figure 20 (Amar, 1974; Brand *et al.*, 1974; Marsland, 1974; Marsland, 1975) show that the variation corresponding to the specified tolerance of ± 5mm/s on the standard of 20mm/s is acceptable.

The relationship between cone resistance and undrained shear strength of a cohesive soil can be expressed as:

$$q_c = N_k \cdot c_u + \sigma_{vo}$$

where σ_{vo} is the total vertical stress, and N_k, the 'cone factor', is analogous to the bearing capacity factor, N_c.

However, N_k is not a constant. It is affected by the method, and reliability, of measurement of c_u, by the shape of the penetrometer tip, by the rate of penetration, by strength anisotropy, and by the macrofabric of the clay and its stiffness ratio (the ratio of shear modulus to undrained shear strength). N_k of OC clays is markedly higher than N_k of NC clays, and it is generally higher when q_c is measured with a mantle cone (M) than with a reference tip (R). Except in some highly-sensitive clays, it is higher than theoretical or experimental values of N_c (usually taken as 9) (Skempton, 1951, Vesic, 1967), because the CPT rate of shearing is some 100 times or more faster than in a field vane test or a laboratory compression test.

40

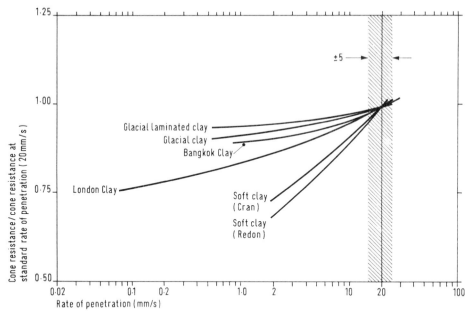

Figure 20 The effect of rate of penetration in clays

6.2   Undrained shear strength of NC clays

6.2.1   Using reference tip

In NC clays, q_c is usually correlated with vane shear strength, s_u, either as measured or as corrected using Bjerrum's (1972) correction derived from back analyses of embankment failures (see Figure 21). The results of an extensive review of cone factors for NC clays by Lunne and Kleven (1981) are shown in Figures 22(a) and 22(b). A cone conforming with the reference test requirements was used, and in almost all cases shear strength was measured with a Geonor field vane. N_k values plotted against plasticity index, I_p, in Figure 22(a), show a variation with I_p, whereas 'corrected' values ($N_k^* = N_k/\mu$, where μ is Bjerrum's correction) in Figure 22(b) are seen to be independent of I_p. The average value in Figure 22(b) is $N_k^* = 15$, the majority of results falling between 11 and 19.

Not all cone factors reported in the literature fall in the range quoted above. Higher values are reported where undrained shear strengths were measured in the laboratory, but these appear to be the result of sample disturbance.. A low value ($N_k = 6.7$; $N_k^* = 8$) was reported by Tümay *et al.* (1982) for recent deltaic deposits of the Mississippi. These were of low sensitivity ($2.2 < S_t < 3.6$) and of high plasticity ($58 < w_L < 77\%$, $37 < I_p < 48\%$). There is no obvious explanation for this low cone factor, but it is possible that the vane shear strengths were increased by the silt and sand lenses present.

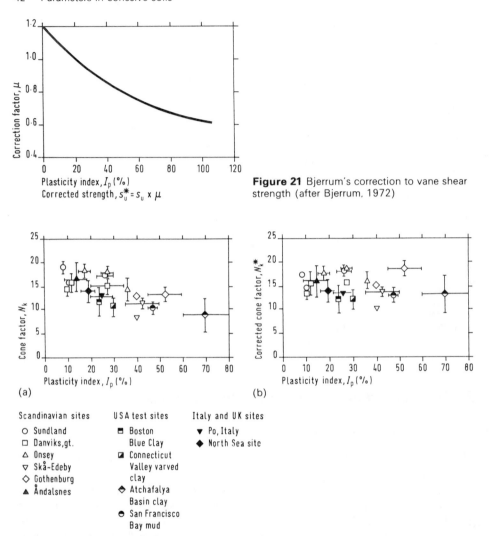

Figure 21 Bjerrum's correction to vane shear strength (after Bjerrum, 1972)

(a)

(b)

Scandinavian sites

○ Sundland
□ Danviks,gt.
△ Onsey
▽ Skå-Edeby
◇ Gothenburg
▲ Åndalsnes

USA test sites

▣ Boston
   Blue Clay
▨ Connecticut
   Valley varved
   clay
◈ Atchafalya
   Basin clay
◒ San Francisco
   Bay mud

Italy and UK sites

▼ Po, Italy
◆ North Sea site

Figure 22 Cone factors for normally-consolidated clays: reference test penetrometer tip (R) (from Lunne and Kleven, 1981) (a) N_R based on vane shear strength as measured. (b) N_R, based on corrected vane shear strength (Bjerrum correction)

Another deviation from N_k^* (average) = 15 was reported by Jamiolkowski *et al.* (1982) for the Porto Tolle site in Italy. They obtained $N_k^* = 11 \pm 3$. This probably results from the very young age of the clay (300 to 400 years).

O'Riordan *et al.* (1982) presented the results of CPTs and vane tests in glacial and post-glacial estuarine and lacustrine clays at three sites in Northern Ireland. Plasticity indices were generally between 10 and 70%. Where I_p was greater than 20%, the average corrected cone factor, N_k^* was 14, ranging from 12 to 18. However, for I_p less than 20%, a large scatter in N_k^* was evident, with many considerably higher values. These higher values of N_k^* were associated with very low vane shear strengths (less than about 5kN/m²), which were lower than indicated by s_u/σ'_{vo} values appropriate to the low plasticity indices (Skempton, 1957). It is not known whether the higher N_k^* values reflect unrepresentative vane shear strengths or whether they result from an unusual structure or condition of the soils.

Clays of high sensitivity present a problem. Roy et al. (1974) reported cone factors which, when expressed as N_k^*, were somewhat lower than 15 (in the range from 11.5 to 13). These were for cemented Champlain Clay, from Eastern Canada, late Pleistocene marine deposits with sensitivities ranging from 16 to well above 50. Ladanyi and Eden (1969) found even lower cone factors (5.5 and 7.5) for two sites in Leda clay, where sensitivities were up to 50 and higher. (A theoretical analysis by Ladanyi (1967) showed that N_k values as low as 5 might be expected for clays of high sensitivity.) On the other hand, Lunne et al. (1976) reported N_k^* values (for Scandinavian clays) of 17.5 to 21 for Drammen clay (sensitivity 50 to 160), 17.5 to 18.5 for Gothenburg clay ($S_t = 15$ to 24) and 15 to 19 for Børrensens Gate clay ($S_t = 15$ to 25).

It can be seen from the foregoing that considerable caution is needed in using the CPT to determine undrained shear strength. For NC clays of low or moderate sensitivity, the choice of a value of N_k^* within the range from 11 to 19 depends on the particular application and the factor of safety to be adopted (see Sections 8.2.1 and 9.4). It is evident that N_k^* does not always fall in this range, and it is therefore desirable that correlations specific to areas and to sites should be developed. Somewhat lower values of N_k^* may apply in very young deposits and recent fills. More sensitive clays need separate consideration, there being insufficient evidence to assign a range of cone factors.

It is possible that some of the differences in N_k values are related to pore-pressure effects (see Section 12.1.1).

6.2.2 Using mantle cone

The use of the Dutch mantle cone (M) or the Dutch friction sleeve penetrometer tip (M) in NC clays gives higher values for cone factor than are obtained with the reference tip (R). This is partly the result of skin friction acting on the mantle (which varies with the sensitivity of the clay) and partly because the pore pressure build-up is smaller with the intermittent action of the mechanical penetrometer than with the continuous action of the electric penetrometer.

Some empirical correlations between cone factor and plasticity index are shown in Figure 23. All the clays are in the soft to soft-to-firm strength range, and they are of moderate sensitivity (average $S_t = 5$; range 3.5 to 7). Strengths were measured by in-situ vane tests, except for the Patras site, where consolidated, undrained, unconfined compression tests were carried out on samples taken with a thin-walled sampler. (The N_k values for Shatt-Al-Arab are higher than those previously presented (Meigh and Corbett, 1969), because they have been adjusted to account for testing at a low penetration rate of 2.5mm/s).

In Figure 23(a), the cone factors are based on vane shear strengths as measured (Anagnostopoulos, 1974, Brand et al., 1974, Meigh, 1969, and Meigh and Corbett,

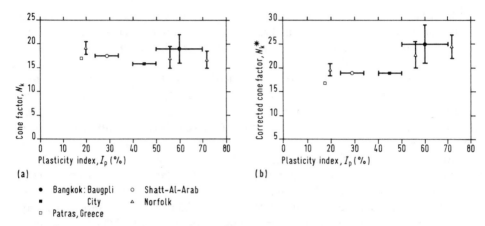

(a)   (b)

● Bangkok: Baugpli   ○ Shatt-Al-Arab
■   City   △ Norfolk
□ Patras, Greece

Figure 23 Cone factor for normally-consolidated clays: mantle cone (a) N_R based on vane shear strength as measured (b) N_R, based on corrected vane shear strength (Bjerrum correction)

1969, and Phan, 1972), whereas, in Figure 23(b), the strengths have been adjusted using the Bjerrum correction. Figure 23(a) shows that the 'uncorrected' cone factor, N_k, is virtually independent of plasticity index, I_p. Correction of N_k, as in Figure 23(b), shows N_k^* varying with I_p, whereas with the reference tip the better correlation is with N_k^*.

Hence, with the mantle cone, N_k should be used to estimate c_u, and c_u should subsequently be corrected for plasticity. The data are limited, but, for the mantle cone in NC clays, they show an average N_k of 17.5, most of the results falling in the range from 15 to 21.

6.3 Undrained shear strength of OC clays

In OC clays, the macrofabric has a marked effect on cone factor, making interpretation in terms of shear strength more dificult and uncertain than with NC clays. The effect of macrofabric varies with the spacing of fissures and other discontinuities. This is illustrated in Figure 24 (Marsland and Quarterman, 1982), where three fissure

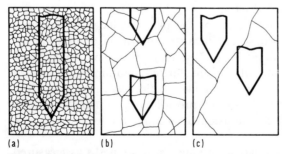

(a)   (b)   (c)

Figure 24 Fissure patterns in overconsolidated clays related to scale of cone penetrometer tip (from Marsland and Quaterman, 1982)

patterns are shown. For the closest spacing (a), the cone resistance reflects the effect of the fissures on the strength of the clay mass. For the intermediate case (b), the cone resistance only partly reflects the effect of the fissures. For case (c) (widely spaced fissures) the strength governing the cone resistance approaches the intact strength of the clay. The effect of scale is further illustrated in Figure 25.

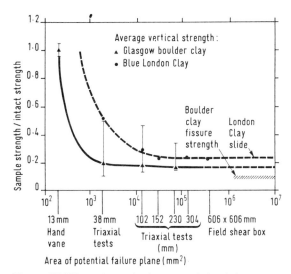

Figure 25 Effect of sample size on undrained shear strength of London Clay and glacial till from Glasgow (from Marsland 1980a)

For determination of N_k values, authors have variously correlated cone resistance of OC clays with field vane shear strength, strength from laboratory compression tests, and strength back analysed from plate-loading tests. Vane strength is unsuitable, because the vane overestimates shear strength in OC clays. Whether laboratory tests are suitable depends on the specimen size in relation to fissure spacing and on the degree of disturbance during sampling and specimen preparation. With wide fissure spacing and low sample disturbance, laboratory test strengths are close to the intact strength of the deposit. With significant sample disturbance, strengths may well fall below the 'mass' strength.

Plate-loading tests give a better basis for comparison provided that the tests are of a sufficient size to take into full account the effect of the fissures. However, they can be misleading, if significant softening takes place before the start of the test, which may happen if tests are carried out below groundwater level, or if drainage occurs during the test.

Figure 26 presents some values of cone factor, based on undrained shear strength back-calculated from plate-loading tests, for marine clays and glacial clays. Zones A, B and C roughly correspond to the fissure patterns in Figure 24. It can be seen that for stiff, fissured marine clays such as London Clay, N_k is in the range from 24 to 30, although somewhat lower values may be found at shallow depth where fissures are

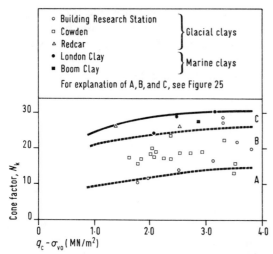

Figure 26 Cone factor for overconsolidated clays, based on undrained shear strength from plate-loading tests (after Marsland and Quarterman, 1982)

more closely spaced. Boom clay is also a fissured marine clay, and a value of $N_k = 27.5$ is shown. This compares with a $N_k = 19$ for Boom clay found by Carpentier (1982), when he based the shear strength on laboratory tests on small specimens.

Glacial clays generally have a macrofabric which has less influence on bulk strength than that of a stiff, fissured clay. The N_k values for glacial clays in Figure 26 average about 18, falling mainly between 14 and 22. The lower N_k values in Zone B, and those in Zone A, apply where glacial clays have a more subdued macrostructure. Semple and Johnson (1979) reported an average N_k of 17 in the northern North Sea and 15 in the central North Sea for glacial clays described as being 'apparently free from structural discontinuities'. Strengths were measured in the laboratory on 55-mm dia. specimens. Some sample disturbance probably occurred, as evidenced by rather low Young's modulus values from the laboratory tests, so that higher N_k values may be appropriate. Nash and Duffin (1982) also based N_k values for glacial clayey till from the north of England on laboratory measurements of shear strength, on 100-mm dia. specimens. Inspection of the samples did not reveal significant discontinuities, although joints at about 300-mm centres were seen in nearby excavations. The average N_k was 19.

Using a tip conforming to reference test requirements, Kjekstad et al. (1978), carried out CPTs with the 'Seacalf' penetrometer, at shallow depth below the sea bed, in non-fissured North Sea glacial clays. They obtained strengths from laboratory compression tests, and found an average N_k of 17.5 within the range 14.5 to 19.5. Down-the-hole penetrometers in the same ground conditions gave N_k values some 15% lower (Lunne and Kleven, 1981.)

Thorburn (1982) found higher N_k values for OC Late Glacial silty clays at Glasgow, $N_k = 25$ on average. However, strengths were measured on 37.5-mm dia. laboratory specimens taken from 150-mm dia. samples from a stationary piston sampler, so that they are probably unrepresentative of the over-all undrained shear strength. A much higher value of N_k was obtained using strengths measured on 37.5-mm dia. specimens taken from 100-mm dia. open-tube samplers, indicating considerable sample disturbance.

Relating the cone resistance of a reference tip to overall shear strength derived from plate-loading tests leads to N_k values for OC clays as follows:

	Average N_k	Range of N_k
Stiff, fissured marine clays	27	24–30
Glacial clays	18	14–22

The choice of the plate-loading test as a reference basis for undrained shear strength has been made in order to get as near as possible to a 'true' *in-situ* value of undrained shear strength. Should it be required to convert the value of c_u so obtained to a laboratory value, two aspects need consideration: sample size and sample disturbance. Figure 25 indicates that the importance of sample size diminishes where the sample diameter is above 100mm (a commonly adopted size) and may be negligible for diameters over about 250mm. The effect of sample disturbance is more difficult to estimate, and judgement based on past experience of the particular formation is required. In general, it can be said that disturbance of OC clays in open-tube sampling increases with increasing undrained shear strength.

The use in OC clays of tips not conforming to the reference test geometry brings additional complications and uncertainties, making interpretation virtually impossible.

6.4   Deformability

6.4.1   Constrained modulus

The constrained modulus, M, for clays can be expressed in terms of a coefficient, α_M, and cone resistance:

$$M = \frac{1}{m_v} = \alpha_M \cdot q_c$$

For NC and lightly OC clays and silts up to firm in consistency (q_c less than about $1.2 \, \mathrm{MN/m^2}$), a first approximation can be obtained by using the α_M values in Table 3. These are applicable for a small stress increment (up to about $100 \, \mathrm{kN/m^2}$) above *in-situ* vertical effective stress.

The data from which the α_M values were obtained were presented by Sanglerat (1979). They are based on cone values from a variety of penetrometers, but mainly from the Dutch mantle cone penetrometer (M). There is a good deal of uncertainty about the ratio between q_c from the mantle cone and q_c from the reference test penetrometer tip (R). On a simple theoretical basis, considering the additional force on the mantle as the product of its area and the remoulded strength of the clay, the ratio can be expected to be between about 1.15 and 1.5 for clays of low or moderate sensitivity $(2 < S_t < 8)$. Limited field evidence suggests a ratio of 1.25, and this has been adopted in arriving at suggested α_M values for the reference tip in Table 3.

Table 3   Coefficient of constrained modulus for normally-consolidated and lightly overconsolidated clays and silts (after Sanglerat, 1979)

Soil	Classification	α_M ($= M/q_c$)	
		Mantle cone (M)	Reference tip (R)
Highly plastic clays and silts	CH, MH	2 to 6	2.5 to 7.5
Clays of intermediate or low plasticity:	CI, CL		
$q_c < 0.7\,MN/m^2$		3 to 8	3.7 to 10
$q_c > 0.7\,MN/m^2$		2 to 5	2.5 to 6.3
Silts of intermediate or low plasticity	MI, ML	3 to 6	3.5 to 7.5
Organic silts	OL	2 to 8	2.5 to 10
Peat and organic clay:	Pt, OH		
$50 < w < 100\%$		1.5 to 4.0	1.9 to 5.0
$100 < w < 200\%$		1.0 to 1.5	1.25 to 1.9
$w > 200\%$		0.4 to 1.0	0.5 to 1.25

Table 3 shows that, for a given q_c, there is a wide range of α_M values. For a better estimate of constrained modulus, it is preferable to use index properties and oedometer test data, but local correlations between q_c and M can be very useful, particularly in assessing variation in compressibility.

For OC clays (q_c greater than about $1.2\,MN/m^2$), Sanglerat's suggested α_M values, for the mantle cone, are given in Table 4. For the reference tip, there are insufficient data to make it possible to recommend α_M values.

6.4.2   Constrained modulus for London Clay

A value of constrained modulus, M, for London Clay can be obtained from:

$$M = \frac{1}{m_v} = 100 \cdot c_u$$

(Skempton, 1951)

Butler (1974) compared observed and calculated settlements of structures founded on London Clay and he found that using the above relationship, together with $E_u = 400\,c_u$, and the Skempton and Bjerrum (1957) method of calculation, there was reasonable agreement:

$$0.88 < \frac{\text{Calculated settlement}}{\text{Observed settlement}} < 1.60$$

However, the c_u values were based on laboratory test data, and in order to arrive at an α_M value from q_c based on c_u from plate-loading tests, it is necessary to allow for the difference arising between these two methods. Evidence from the London Clay at

Table 4   Coefficient of constrained modulus for overconsolidated clays and silts (mantle cone) (after Sangerat, 1979)

Soil	Classification	α_M $(= M/q_c)$	
		$1.2 < q_c < 2.0\,\text{MN/m}^2$	$q_c > 2.0\,\text{MN/m}^2$
Highly plastic silts and clays	MH, CH	2 to 6	—
Clays of intermediate or low plasticity	CI, CL	2 to 5	1 to 2.5
Silts of low or intermediate plasticity	MI, ML	3 to 6	1 to 3

Chelsea (Marsland, 1971b), and at Hendon (Marsland, 1974) shows that the difference increases with increasing depth:

At Hendon $c_{u(lab)} \simeq c_{u(PLT)} (1 + 0.02Z)$

At Chelsea $c_{u(lab)} \simeq c_{u(PLT)} (1 + 0.06Z)$

where Z is the depth in metres below the upper surface of the London Clay. Averaging these:

so that: $c_{u(lab)} \simeq c_{u(PLT)} (1 + 0.04Z)$ $\qquad$ (3)

$M \simeq 100 (1 + 0.04Z) . c_{u(PLT)}$ $\qquad$ (4)

At shallow depth, this is equivalent to an α_M value between 3 and 4, compared with Sanglerat's range from 2 and 6 (Table 4).

Other London Clay sites may differ, depending on local variations in geological history, the orientation, size and nature of discontinuities, and other fabric features. However, Equation (4) should give a reasonable approximation.

6.4.3   Undrained Young's modulus

There is even greater difficulty in assessing a value of undrained Young's modulus, E_u, from q_c values. There are insufficient data available to make a direct correlation, and it is recommended that c_u should first be derived, using the correlations given in Section 6.2, then E_u estimated, as a rough order of value from one of the available correlations between E_u and c_u. Many such correlations have been proposed. A useful one (Duncan and Buchignani, 1976), which has been determined from field measurements, is shown in Figure 27. E_u decreases with increasing shear stress level. An approximate basis (Ladd *et al.* 1977) for estimating such a reduction for NC clays is given in Figure 28. This is based on the assumption that existing correlations between E_u and c_u relate to a factor of safety of about 4 (i.e. $\tau/c_u \simeq 0.25$).

There are some data concerning UK glacial clays*, shown in Figure 29, both in terms of E_u/c_u and E_u/q_c. Values were derived from plate-loading tests and CPTs

---

* Weltman and Healy (1978) derived values of $E_{u(50)}$ from the analysis of a large number of loading tests on piles in glacial till. The ratio of E_u/c_u was found to increase strongly with increasing c_u, more strongly than results from decreasing OCR. However, strengths were determined from laboratory tests obtained during standard ground investigations, and in such tests disturbance increases markedly with increasing *in-situ* shear strength.

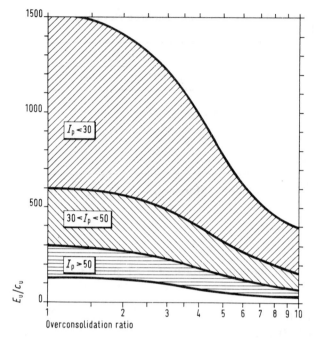

Figure 27 Ratio of undrained Young's modulus to shear strength against overconsolidation ratio for clays (after Duncan and Buchignani, 1976)

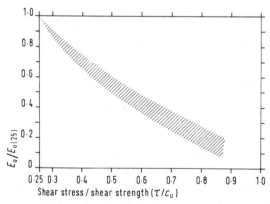

Figure 28 Reduction of undrained Young's modulus of normally-consolidated clays with increasing shear stress level (based on data from Ladd *et al.*, 1977)

undertaken by the Building Research Station (Marsland, 1975, 1977, 1980). E_u values are first loading secant moduli to 50% of failure stress, $E_{u(50)}$. It is found that $E_{u(25)}$ and E_{uo} (initial tangent modulus) values are of the order of 50% higher.

Figures 27 and 29 show that E_u/c_u and E_u/q_c vary with OCR (i.e. the ratio of maximum past pressure and present vertical effective stress). In some cases, the OCR can be assessed from geological evidence. Alternatively, an approximate value can be

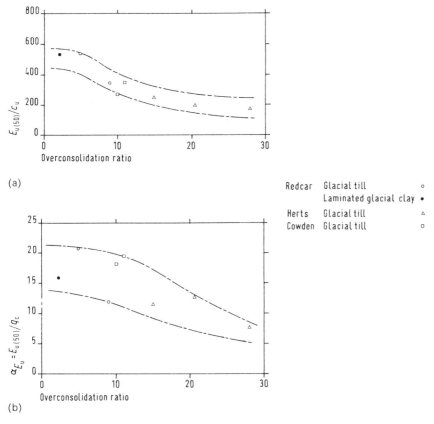

(a)

Redcar	Glacial till	o
	Laminated glacial clay	•
Herts	Glacial till	△
Cowden	Glacial till	□

(b)

Figure 29 Undrained Young's modulus for UK glacial clays (a) Ratio of Young's modulus to undrained strength (c_u from plate loading tests). (b) Ratio of Young's modulus to cone resistance

obtained from Figure 30, using the ratio of c_u/σ'_{vo} to $(c_u/\sigma'_{vo})_{NC}$ where $(c_u/\sigma'_{vo})_{NC}$ is the ratio of undrained shear strength to vertical effective stress which applies to NC soil. This can be derived (Skempton, 1957) from:

$$c_u/\sigma'_{vo} = 0.11 + 0.0037 I_p$$

where I_p is the Plasticity Index (%). If I_p is not known, an average value of $c_u/\sigma'_{vo} = 0.3$ can be used.

A study by Mayne (1980) of documented values of the ratio of normalised undrained shear strength of OC clays to that of NC clays showed a wider range of values than that given in Figure 30. This reference should be studied if a closer estimate of the ratio is required.

6.4.4   Undrained Young's modulus of London Clay

There are insufficient data to produce plots similar to Figure 29 for London Clay.

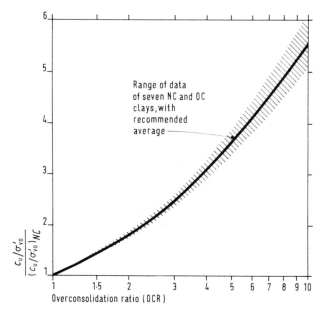

Range of data
of seven NC and OC
clays, with
recommended
average

$\dfrac{c_u/\sigma'_{vo}}{(c_u/\sigma'_{vo})_{NC}}$

Overconsolidation ratio (OCR)

Figure 30 Basis for estimation of overconsolidation ratio (from Schmertmann, 1978)

Results from one site (Marsland, 1971(a), 1974) suggest an average E_u/q_c value of 23.5 (range 20 to 27). However, as mentioned in Section 6.4.2, Butler (1975) found a good correlation between observed and calculated settlements of structures on London Clay using:

$$E_u = 400c_{u(lab)}$$

In terms of undrained shear strength derived from plate-loading tests, this gives, approximately:

$$E_u \simeq 400(1 + 0.04Z) \cdot c_{u(PLT)} \tag{5}$$

where Z is the depth in metres below the top of the London Clay.

Using the N_k values for stiff, fissured marine clays given on page 47, at shallow depth this is equivalent to a value of

$$\alpha_{E_u} E_u/q_c = 15 \pm 2$$

This is much lower than the Marsland value for one site quoted above, namely 23.5 ± 3.5. Because Butler's relationship is derived from observed settlements, it is preferable to use Equation (5), together with $N_k = 27 \pm 3$, in evaluating E_u for London Clay, rather than the direct relationship between E_u and q_c.

A higher value of E_u applies in problems such as movements of the walls of excavations in London Clay, where strains are of the order of 0.05% compared with strains of some 0.1 to 0.2% associated with settlements. Simpson *et al.* (1979) suggest, for London Clay:

$$E_u = 1000\,c_{u(lab)}$$

6.4.5   Drained Young's modulus

There are no data available relating drained Young's modulus, E' to q_c from the CPT. In theory, E' can be derived from E_u, using appropriate values of Poisson's ratio, v, but this is not borne out in practice.

6.4.6   Drained Young's modulus of London Clay

For London Clay, Butler (1975) suggests that the vertical drained Young's modulus can be taken as:

$$E'_v = 130c_{u(lab)} = 0.325E_u$$

In terms of undrained shear strength from plate-loading tests, this gives:

$$E'_v = 130\,(1 + 0.04Z).c_{u(PLT)}$$

where Z is the depth in metres below the top of the London Clay.
At shallow depth, this is equivalent to a value of

$$\alpha_{E'_v} = E'_v/q_c = 4.8 \pm 0.5$$

6.5   Synopsis: parameters in cohesive soils

6.5.1   Undrained shear strength

(a) Normally-consolidated clays
If reference tip (R) is used:
$\quad N_k^* = 15 \pm 4$ (Figure 22b)

where $q_c = N_k^* s_u^* + \sigma_{vo}$

and s_u^* is vane shear strength with the Bjerrum correction (Figure 21)

Refer to Section 6.2.
Refer to Section 6.2.1.

For very young deposits and recent fills, the average value of N_k^* may be less than 15.

Considerably higher values and wide scatter have been reported for a case of low I_p ($<20\%$) and very low s_u ($<5\,kN/m^2$). Some clays of high sensitivity have shown very low N_k^* values.

If mantle cone (M) used:

Refer to Section 6.2.2.

$\quad N_k = 17.5$ (range: 15 to 21)
(Figure 23a)

where $q_c = N_k s_u + \sigma_{vo}$
and s_u is uncorrected vane shear strength.

(b) Overconsolidated clays
For reference tip (R)

$$q_c = N_k c_u + \sigma_{vo}$$

In stiff, fissured clays

$$N_k = 27 \pm 3$$

In UK glacial clays

$$N_k = 18 \pm 4$$

6.5.2 Deformability

(a) Constrained modulus, M

$$M = \frac{1}{m_v} = \alpha_M \, q_c$$

For normally-consolidated clays, Table 3.
For overconsolidated clays, Table 4.
For London Clay

$$M \simeq 100 \, (1 + 0.04Z) c_{u(PLT)}$$

(b) Undrained Young's Modulus
General approach: estimate E_u from Figure 27
(or similar correlation), using values of c_u
(Section 6.2) and OCR (Figure 30).

For UK glacial clays, use Figure 29
For London Clay

$$E_u \simeq 400 \, (1 + 0.0.04Z) c_{u(PLT)}$$

(c) Drained Young's Modulus
For London Clay
$$E_v' \simeq 130 \, (1 + 0.04Z) c_u(PLT)$$

Refer to Section 6.3.

c_u is that from plate-loading tests.
N_k varies with macro fabric (see Figures 24, 25 and 26).

Refer to Section 6.4.

Refer to Section 6.4.1.

Applicable for stress increment of up to about $100 \, kN/m^2$.

For a better estimate of M, use index properties and oedometer test data.

Refer to Section 6.4.2.

Refer to Section 6.4.3.
E_u decreases with increasing shear stress level (see Figure 28 for normally-consolidated clays).

$$c_u/\sigma_{vo}' = 0.11 + 0.0037 Ip \simeq 0.03$$

Refer to Section 6.4.4.
For movements of walls of excavations:
$E_u = 1000 c_{u(lab)}$
At shallow depth in London Clay
$E_u/c_c = 15 \pm 2$

Refer to Section 6.4.6.
At shallow depth in London Clay
$E_v'/q_c = 4.8 \pm 0.5$

7 Parameters in other materials

Sections 4, 5 and 6 relate primarily to sands and clays without a significant carbonate content. Silts have not been specifically dealt with, but in general terms coarse silts can be treated as sands, and fine, cohesive silts as clays. This Section deals with a variety of other materials.

7.1 Calcareous soils

Calcareous soils vary widely in their characteristics, depending on their origin, carbonate content, form of constituents and degree of cementation (Demars and Chaney, 1982). Frequently, they contain shells or shell fragments and other fragile remains of calcareous marine life, which influence compressibility and other engineering parameters. Many calcareous soils are difficult to sample and sensitive to sample disturbance. Where cementation is present, it can be destroyed by drive sampling. *In-situ* testing is therefore particularly appropriate, but there is little experience of cone penetration testing in these materials. Although some authors found that relationships applying to more conventional materials can be adopted for certain calcareous soils, others have not.

Beringen *et al.* (1982) investigated calcareous sands and silts in the Rankine field offshore from Western Australia, and they found a close correlation between cone resistance (reference tip) and cementation. This is a site-specific correlation. They recommend that a modification of the system proposed by Clark and Walker (1977) should be used for the classification of calcareous soils containing 90% or more of carbonate, as shown in Figure 31. For these Western Australian carbonate soils, Beringen *et al.* found a reasonable correlation between friction ratio and cone resistance (representing degree of cementation), cemented conditions being characterised by high cone resistance and relatively low friction ratios (see Figure 32).

In Figure 33, friction ratios for offshore Bombay clays with a carbonate content varying between 40 and 70% are compared with friction ratios quoted by De Ruiter (1975) for North Sea soils. Typically, for North Sea soils, the friction ratio is less than 3% for quartz sands and varies between about 2 and 8% for conventional clays. It can be seen that similar values apply to the Bombay calcareous soils. Beringen *et al.*

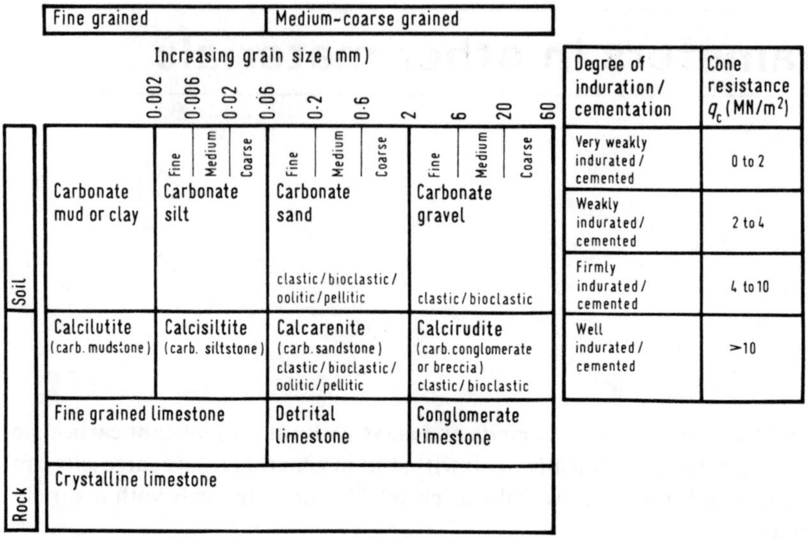

Fine grained
Medium-coarse grained
Increasing grain size (mm)
0·002
0·006
0·02
0·06
0·2
0·6
2
6
20
60
Soil
Rock
Fine
Medium
Coarse
Carbonate mud or clay
Carbonate silt
Carbonate sand
clastic/bioclastic/oolitic/pellitic
Carbonate gravel
clastic/bioclastic
Calcilutite (carb. mudstone)
Calcisiltite (carb. siltstone)
Calcarenite (carb. sandstone) clastic/bioclastic/oolitic/pellitic
Calcirudite (carb. conglomerate or breccia) clastic/bioclastic
Fine grained limestone
Detrital limestone
Conglomerate limestone
Crystalline limestone
Degree of induration/cementation
Cone resistance q_c (MN/m^2)
Very weakly indurated/cemented
0 to 2
Weakly indurated/cemented
2 to 4
Firmly indurated/cemented
4 to 10
Well indurated/cemented
>10

Figure 31 Classification of carbonate sediments, 90% to 100% carbonate (after Clark and Walker, 1977, as modified by Beringen, Kolk and Windle, 1982)

Line of separation between sands and clays for North Sea soils
SANDS
CLAYS
Well cemented
Firmly cemented
Weakly cemented
Very weakly cemented
0·5
0·4
Cone resistance, q_c (MN/m^2)
Friction ratio, R_f (%)
40
20
10
8
6
4
2
1
0
1
2
3
4
5
6
7
8

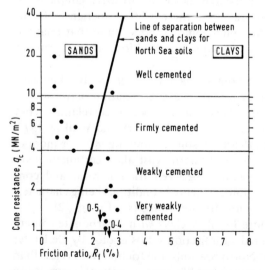

Figure 32 Cone resistance and friction ratio for calcareous sands and silts offshore Western Australia (after Beringen, Kolk and Windle, 1982)

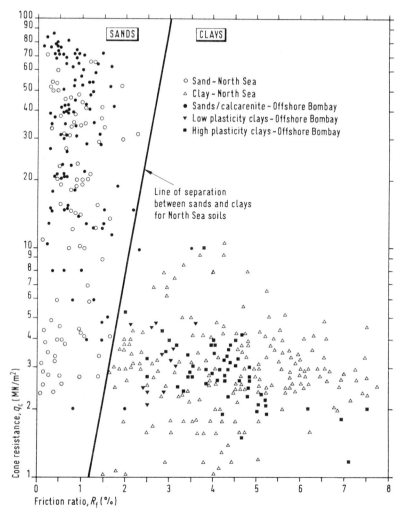

Figure 33 Cone resistance and friction ratio for calcareous soils offshore Bombay compared with North Sea soils (from Beringen, Kolk and Windle, 1982)

(1982) found that the values of cone factor, N_k, for the Bombay calcareous soils were also similar to North Sea values.

7.2 Chalk

Chalk is disturbed by driven-tube sampling to the extent that samples are useless for laboratory testing and have little value for identification. Rotary coring is also often unsatisfactory, particularly in the Upper Chalk, where flints abound. The CPT is therefore of considerable interest, although it also suffers where flints are present.

Chalk has been the subject of much research, most notably in the Middle Chalk at Mundford, Norfolk (Ward *et al.*, 1968, Burland and Lord, 1969). At this site, the deformation characteristics of the chalk were investigated with a series of plate-loading tests in accessible augered boreholes and the full-scale loading of a 18.3-m dia. water tank.

Detailed examination within the boreholes led to the development of a Chalk Grade classification, which was extended by Wakeling (1969) to include structureless chalk ('putty' chalk), Grade VI, as shown in the left-hand side of Table 5. Piling in chalk was the subject of a comprehensive review by Hobbs and Healy (1979). The design procedures recommended in that publication are based on the SPT, together with careful description of the chalk. It is particularly useful in its guidance on such description, and its review of the various kinds of chalk.

Table 5   Chalk grades related to CPT values (after Power, 1982)

Grade	Brief description	q_c (MN/m²)	R_f (%)
VI	Extremely soft structureless chalk containing small lumps of intact chalk	< 5	—
V	Structureless remoulded chalk containing lumps of intact chalk	5–15	0.75–1.0
IV	Rubbly partly-weathered chalk with bedding and jointing. Joints 10 to 60mm apart, open to 20mm, and often infilled with soft remoulded chalk and fragments	5–15	1.0–1.25
III	Rubbly to blocky unweathered chalk. Joints 60 to 200mm apart, open to 3mm, and sometimes infilled with fragments	15–20	1.25–1.50
II	Blocky medium-hard chalk. Joints more than 200mm apart and closed	> 20	1.5–2.0
I	As for Grade II, but hard and brittle	No penetration	

A typical CPT profile from the Mundford site is presented in Figure 34 (Power, 1982). The sharp peaks in the profile are thought to be not entirely the result of flints, but rather the manner in which penetration resistance builds up, then is followed by grain crushing and/or closure of fissures, together with the possible effects of variability in density, degree of cementation and jointing and fissuring.

A scheme for the interpretation of CPT results in chalk using the Dutch friction sleeve penetrometer tip (M) is presented by Searle (1979), as shown in Figure 35. For the reference test penetrometer tip (R), Power (1982) relates chalk grade to q_c and R_f, as shown in the right-hand side of Table 5.

For Mundford and for a number of other sites in the Middle and Upper Chalk, Power investigated the relationship between q_c and SPT N value. The results are presented in Figure 36. The data include points for both reference tips and mechanical tips not conforming to the reference test tip. According to Power, most of the results for the reference tip fall in the range $0.3N < q_c < 0.7N$, with a median value, $q_c = 0.5N$ MN/m². Many values fall below $q_c = 0.3N$. Power points out that this relationship may not apply in other areas (e.g. Suffolk) where the Upper Chalk is weaker, irrespective of the degree of weathering, and where preliminary correlations plot consistently between the median and upper-bound lines.

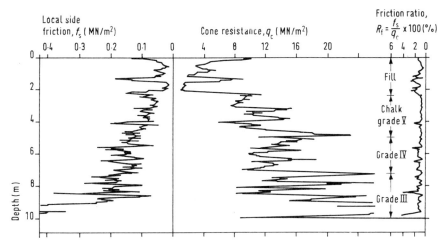

Figure 34 Typical cone penetration test profile in Middle Chalk at Mundford, Norfolk (after Power, 1982)

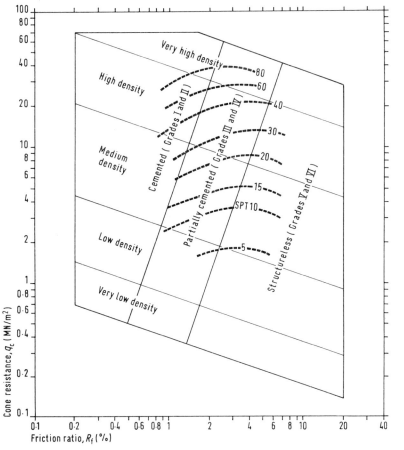

Figure 35 Proposed interpretation chart for chalk: Dutch friction sleeve penetrometer (M) (from Searle, 1979)

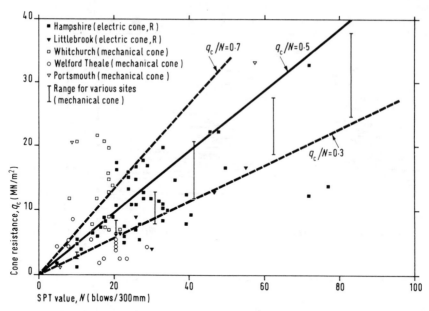

Figure 36 Relationship between cone resistance and standard penetration test value for Middle and Upper Chalk (from Power, 1982)

For the Mundford site, Power also investigated the relationship between Young's modulus, E and q_c. A tentative relationship is given in Figure 37. Care should be exercised in extrapolating from one area and chalk type to another. Ideally, specific correlations need to be developed for each site. Furthermore, to facilitate compari-

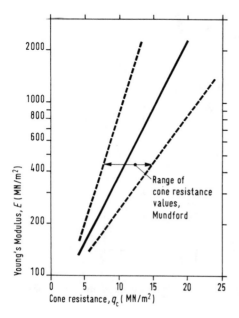

Figure 37 Relationship between cone resistance and Young's modulus for Middle Chalk at Mundford, Norfolk (from Power, 1982)

sons between sites, it is essential that CPT results should be accompanied by careful descriptions of the chalk.

7.3 Made ground, including tailings

In made ground placed under controlled conditions, such as hydraulic fills, the CPT is generally suitable for examining variability and assessing engineering parameters. This is clearly not so where rock fill or numerous cobbles or boulders predominate. The CPT is seldom suitable for the exploration of uncontrolled made ground (e.g. domestic refuse or demolition debris). In such materials, investigation is best made by means of trial pits or trenches.

A special category of made ground is found in mine tailings deposits. Many tailings dams were constructed without detailed and planned engineering design, and subsequently require investigation, particularly where they may be subjected to earthquake shaking. In such cases the CPT might be useful in delineating thin zones with a high fines content. It is also necessary to determine pore-pressure conditions within the tailings and within and below the tailings dam, also to evaluate engineering parameters (particularly the angle of shearing resistance) in terms of effective stress. The piezocone may be useful for this purpose (see Sections 11.5 and 12.2.3).

8 Spread footings

It is again emphasised that caution is required when applying empirically-derived parameters in design.

8.1 Footings and rafts on sand

As in all foundation design, it is necessary to consider both safe bearing capacity (i.e. the ultimate bearing capacity divided by a suitable factor of safety) and allowable bearing capacity related to tolerable settlements.

8.1.1 Safe bearing capacity

Ultimate bearing capacity (and hence safe bearing capacity) can be calculated using the standard bearing capacity formula (Terzaghi, 1943) and bearing capacity factors (Terzaghi, 1943, for shallow foundations; Meyerhof, 1951, for deep foundations). Values of ϕ can be obtained from Figure 15 (in Section 5.3) modified if necessary in the light of the accompanying discussion.

Except for narrow foundations on relatively loose sand, bearing capacity is seldom a problem, and the selection of an allowable bearing pressure is governed by settlement considerations.

8.1.2 Settlement

Generally, the direct use of q_c in settlement calculations for foundations on sand is preferable to methods in which q_c is first converted to SPT blow count, N. However, there is a quick check method via SPT (Burland *et al.* 1977) which is often useful to indicate the probable extent of a settlement problem. This method uses the data presented in Figure 38, which are from site records where ground conditions and settlements were known. Upper limits are drawn for both dense sand ($N > 30$) and medium dense sand ($10 < N < 30$).

In each case, the authors suggest that probable settlement can be taken as equal to half the upper limit value, and that maximum settlement does not normally exceed about 1.5 times the probable value. The upper limit for loose sands ($N < 10$) is regarded as tentative. Much of the data in the upper zone relates to very loose, slightly silty, organic sands, and the authors suggest that the upper limit values could be useful in the preliminary assessment of settlement of structures such as large oil tanks on loose sand. To use Figure 38, a D_{50} value has to be known or assumed with N determined from q_c using Figure 19 (in Section 5.5).

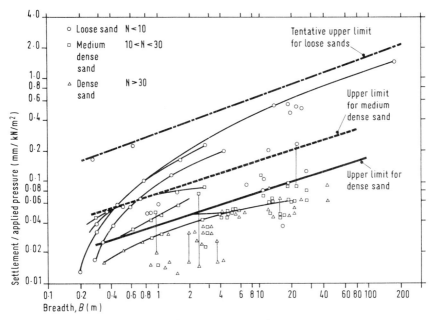

Figure 38 Observed settlement of footings on sand of various relative densities (after Burland, Broms and de Mello, 1977)

Another method which has considerable merit, although it is indirect, is that of Burland and Burbidge (1985). This method is derived from an extensive review of case histories, mostly based on SPT but including some where the CPT was used.

A rapid conservative estimate of settlement of a footing on sand can be directly obtained from q_c, using the relationship proposed by Meyerhof (1974):

$$s = \frac{p_n B}{2\bar{q}_c}$$

where p_n is the net applied loading, $\bar{q}_c$ is the average value of q_c over a depth equal to the footing width, B, and s is settlement. This is roughly equivalent to using a Young's modulus, $E = 1.5q_c$, compared with Schmertmann's value of $2.5q_c$ (see Section 5.4.2).

A method which has been much used, but has now been superseded, is that of De Beer and Martens (1957). It is based on the Terzaghi–Buisman formula:

$$s = \int_0^H \left[\frac{2.3}{C} \log_{10} \left(\frac{\sigma'_{vo} + \Delta\sigma_v}{\sigma'_{vo}} \right) \right] \Delta H$$

where the 'constant of compressibility', C, is given by:

$$C = \frac{3}{2} \frac{q_c}{\Delta \sigma_v'}$$

This method gives an often considerable overestimate of settlement.

For direct use of CPT values in calculating settlements of footings on sand, the Schmertmann method (1970), as modified by Schmertmann, Hartman and Brown (1978) (and fully described in Schmertmann, 1978), is probably the best available. In this method (see Figure 39), the sand is divided into a number of layers, n, of equal thickness, Δz, down to a depth below the base of the footings equal to $2B$ for a square footing and $4B$ for a long footing ($L \geqslant 10B$). Strain within each layer is taken as $I_z \Delta_p / E$ where I_z is a 'strain influence factor' (Figure 39(a)).
Thus

$$s = C_1 C_2 \Delta_p \sum_1^n \frac{I_z}{x q_c} . \Delta z$$

where C_1 and C_2 are corrections for depth of embedment and creep, respectively; Δ_p, described as a net foundation pressure increase, is equal to the applied pressure, minus the *effective* overburden pressure at foundation level, σ_{vo}'; and $E = x q_c$.

The factor x is analogous to α in the expression $E = \alpha q_c$, but includes shape factors.

For a square footing, $x = 2.5$
For a long footing ($L \geqslant 10B$), $x = 3.5$

In the calculation, the strain distribution diagram shown in Figure 39(a) is redrawn to correspond to the peak value of I_z obtained from $I_{zp} = 0.5 + 0.1 (\Delta_p / \sigma_{vp}')^{0.5}$. The sand is then divided into a convenient number of layers, n (say 8 for $L/B = 1$, 8 or 16 for $L/B = 10$) and a value of q_c is assigned to each layer. The value of $I_z / x q_c$ is then determined for each layer, where I_z at the mid depth of the layer is taken from the redrawn Figure 39(a). The values of $I_z / x q_c$ are then summed, and the sum is multiplied by Δp, C_1 and C_2. (For $1 < L/B < 10$, the results can be interpolated between the $L/B = 1$ and $L/B = 10$ cases).

The depth of embedment correction is:

$$C_1 = 1 - 0.5 (\sigma_{vo}' / \Delta_p)$$

with a minimum value of 0.5.

The creep correction is:

$$C_2 = 1 + 0.2 \log_{10} (10t)$$

where t = time in years from load application.

The values of x quoted above are applicable for NC sands, and they are based on a load increment from 100 to 300kN/m². It is probable therefore that somewhat higher x values may be appropriate for loose sands and somewhat lower values for very dense sands.

As discussed in Section 5.4.2, Young's modulus, E, for OC sands is very

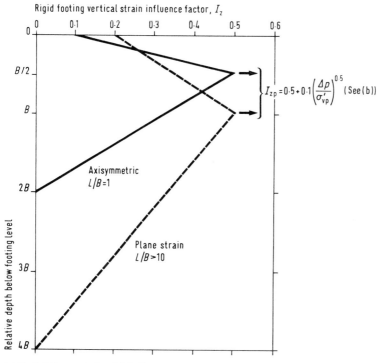

(a) Strain influence factor distributions

Note: In calculation, the diagram
      is redrawn to
      correspond to I_{zp}

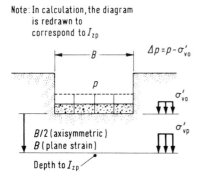

$\Delta p = p - \sigma'_{vo}$

(b) Explanation of pressure terms
    in equation for (a)

Figure 39 Strain influence factor diagrams (after Schmertmann, Hartman and Brown, 1978)

considerably higher, but it is suggested that values not more than twice those for NC sands should be adopted. Thus:

For $L/B = 1$, $x_{oc} = 5$
For $L/B = 10$, $x_{oc} = 7$

8.2 Footings and rafts on clay

As with footings on sand, it is necessary to consider both safe and allowable bearing capacity.

8.2.1 Safe bearing capacity

Ultimate (and hence safe) bearing capacity can be calculated from undrained shear strength, c_u, using standard formulae (e.g. Skempton, 1951). This requires values of N_k for use in the equation:

$$q_c = N_k \cdot c_u + \sigma_{vo}$$

The value of N_k (or N_k^*) to be adopted is influenced by the factor of safety to be used, and other considerations. In NC clays, a probable value for N_k^* is 15, and a more cautious value 19. This may be very conservative in the case of clays of high sensitivity. The comments in Section 6.2 should also be studied.

If a mantle cone has been used, the corresponding values of N_k are 17.5 and 21, but these give uncorrected values of c_u, and to obtain a corrected value it is necessary to apply the Bjerrum correction (see Figure 21, Section 6.2), which requires a value or estimated value of plasticity index.

For OC clays (reference tip), values of N_k are:

Clay type	Probable value of N_k	More cautious value of N_k
Stiff-fissured marine	27	30
UK glacial	18	22

Resulting values of c_u correspond to values determined from plate-loading tests.

8.2.2 Settlement

Two methods are discussed below:

1. the method of Skempton and Bjerrum (1957), which combines immediate settlement evaluated by elastic theory and consolidation settlement from a modification of Terzaghi's theory of consolidation
2. a method using linear elastic theory to calculate total settlement.

Skempton and Bjerrum's method is presented in most standard textbooks on soil mechanics. It requires a value of undrained Young's modulus, E_u, and constrained modulus, $M = 1/m_v$.

As discussed in Section 6.4.3, E_u, is obtained by first determining c_u from q_c, using the correlations given in Section 6.2, then using one of the correlations available between E_u and c_u (e.g. that given in Figure 27, page 50). For normal structural foundations, values of E_u/c_u from Figure 27 require no adjustment. But if a low factor of safety is used (e.g. for structures such as oil storage tanks) E_u/c_u should be reduced, as indicated in Figure 28. Guidance on selection of values of M can be obtained from Tables 3 and 4 in Section 6.4.1.

If the foundation is on London Clay, parameters can be adopted as follows:

$$M \simeq 100(1+0.04Z).c_{u(PLT)}$$
$$E_u \simeq 400(1+0.04Z).c_{u(PLT)}$$

For UK glacial clays, values of E_u/c_u and E_u/q_c can be obtained from Figure 29 (in Section 6.4.3). However, these values should be used with caution in view of the limited data on which they are based.

As an alternative to the Skempton and Bjerrum method, a linear elastic calculation can be used, with the vertical drained Young's modulus, E_v'. A method which allows for E_v' increasing linearly with depth was presented by Butler (1975) and extended by Meigh (1976). Unfortunately, values of E_v'/c_u and E_v'/q_c are not generally available. For London Clay, Butler (1975) showed that good correlation between calculated and observed settlements is given by

$$E_v' = 0.325 E_u$$

and from Equation (5), page 52

$$E_v \simeq 130 \,(1+0.04Z).c_{u(PLT)}$$

As a first approximation, it is reasonable to use $E_v' = 0.325 E_u$ for other UK stiff, fissured marine clays.

8.3   Synopsis: spread footings

8.3.1   Footings and rafts on sand

Refer to Section 8.1

(a) *Safe bearing capacity*

Obtain values of ϕ' from figure 15, page 31. Refer to Section 8.1.1.

Use the standard bearing capacity formulae and bearing capacity factors:
   for shallow foundations: Terzaghi (1943)
   for deep foundations: Meyerhof (1951).

(b) *Settlement*

Refer to Section 8.1.2.

For a quick check, convert q_c to N using Figure 19, and refer to Figure 38.

Probable settlement can be taken as half the upper limit.

The upper limit for loose sands ($N < 10$) is tentative, relating to very loose, slightly silty, organic sands and probably applicable to large oil tanks, etc. on such soils.

For a more detailed analysis, again convert q_c to N and use Burland and Burbidge (1985).

To estimate settlement from CPT results either

$$s \simeq \frac{p_n B}{2\bar{q}_c}$$

where $\bar{q}_c$ is the average cone resistance over a depth equal to the footing width, B

or

$$s = C_1 . C_2 . \Delta_p \sum_1^n \frac{I_z}{x\,q_c} . \Delta z$$

Refer to page 63.

See pages 64 and 65, and Figure 39.

8.3.2   Footings and rafts on clay

Refer to Section 8.2.

(a) *Safe bearing capacity*

Refer to Section 8.2.1.

Use standard formulae (e.g. Skempton (1951)) with c_u determined from q_c through the relevant cone factor N_k or N_k^*. Probable values

for normally-consolidated clay
    with reference tip $N_k^* = 15$
    with mantle cone $N_k\ \ = 17.5$

More cautious value $N_k^* = 19$.
More cautious value $N_k = 21$.

for overconsolidated clays with reference tip
    stiff, fissured marine clays $N_k = 27$
    UK glacial clays $N_k = 18$

More cautious value $N_k = 30$.
More cautious value $N_k = 22$.

(b) *Settlement*

Refer to Section 8.2.2.

Either use the method of Skempton and Bjerrum (1957), calculating:
and immediate settlement, using E_u
    consolidation settlement, using $M = 1/m_v$
or use Linear Elastic method based on E_v' (Butler (1975), Meigh (1976)).

Refer to Section 6.4.3.
Refer to Section 6.4.1.

For London Clay, $E_v' = 0.325 E_u$.

As a first approximation, it is reasonable to use this relationship for other UK stiff fissured clays.

9 Piles

The ultimate bearing capacity of a pile, Q, is the sum of the ultimate end-bearing capacity, Q_b, and the ultimate shaft resistance, Q_s. Safe bearing capacity, Q_{safe} is then calculated by applying a factor of safety to Q or separate factors of safety to the components Q_b and Q_s. Q_b dominates in sands, and Q_s in clays (except for the case of short piles with an enlarged base). Allowable bearing capacity depends on the settlement which can be tolerated.

9.1 Driven piles in sand, bearing capacity

The methods described below are applicable to mainly quartz sands. They are not directly applicable to gravels, because the bearing capacity of a pile in gravel is less than indicated by cone resistance. The methods have only limited applicability in carbonate sands (see Section 9.5).

9.1.1 Ultimate end bearing

In a uniform deposit of sand, below a certain depth a parallel-sided displacement pile achieves an ultimate bearing capacity equal to the cone resistance.

$$\left(Q_b = q_p \cdot A_b, \text{ where } q_p = \frac{q_{c1} + q_{c2}}{2} \right)$$

The depth below which this occurs is known as the critical depth. It varies with soil stiffness (and possibly with pile diameter), and it ranges between 4 and 20 pile diameters, the critical depth increasing with increasing soil stiffness. A typical value of 8 is often adopted.

Sand deposits are seldom uniform, and, in practice, it is necessary to derive a composite q_c value, q_p, to take account of the variation of q_c above and below the pile toe, and a procedure for doing this (Heijnen, 1974) is shown in Figure 40. In evaluating q_{c2}, trials are made with a number of depths, below pile toe, between $0.7d$ and $4.0d$, and the lowest resulting q_{c2} is adopted. Typical Dutch practice in assessing q_p is to limit the value of q_c used (normally to 30MN/m^2), and to limit the ultimate end-bearing capacity to a value not exceeding 15MN/m^2, which depends on OCR as shown in Figure 41 (te Kamp, 1977). Some further reduction may be required if weaker layers exist between $4d$ and $10d$ below pile toe level.

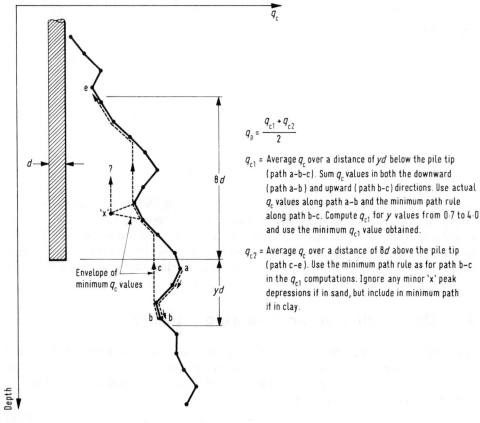

$$q_p = \frac{q_{c1} + q_{c2}}{2}$$

q_{c1} = Average q_c over a distance of yd below the pile tip (path a–b–c). Sum q_c values in both the downward (path a–b) and upward (path b–c) directions. Use actual q_c values along path a–b and the minimum path rule along path b–c. Compute q_{c1} for y values from 0·7 to 4·0 and use the minimum q_{c1} value obtained.

q_{c2} = Average q_c over a distance of $8d$ above the pile tip (path c–e). Use the minimum path rule as for path b–c in the q_{c1} computations. Ignore any minor 'x' peak depressions if in sand, but include in minimum path if in clay.

Figure 40 Procedure for determining composite cone penetration test value in evaluation of pile end-bearing capacity (after Schmertmann, 1978 and Heijnen, 1974)

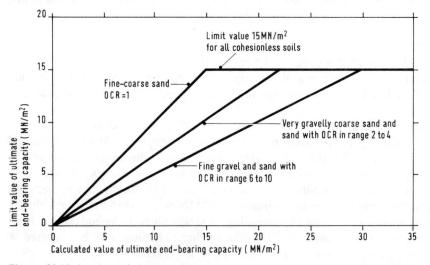

Figure 41 Limit values of ultimate pile end-bearing capacity (after te Kamp, 1977)

An alternative procedure is proposed by Meyerhof (1976) for assessing capacity above the critical depth. This is less complex to apply than the Dutch procedure, and it does not involve the use of the 15-MN/m^2 cut-off limit. However, a reduction factor has to be applied for piles of diameter greater than 0.5 m, as subsequently recommended by Meyerhof (1983) (see Section 9.2).

9.1.2 Ultimate shaft resistance from local side friction

Shaft resistance can be calculated from values of local side friction, f_s. However, measurement of cone resistance is often more accurate and more easily interpreted than that of f_s, so that shaft resistance is frequently based on q_c rather than on f_s.

Shaft resistance for a driven pile in sand depends not only on the properties of the sand, but also on the extent to which the density of the sand is modified by pile driving. The ultimate shaft resistance is given by:

$$Q_s = \sum_0^L q_s \pi d\Delta L = S_1 \sum_0^L f_s \pi d\Delta L \qquad (6)$$

where q_s = unit ultimate shaft resistance
$\quad\quad L$ = length of pile in the sand

and where the value of S_1, which depends on the type of pile, is taken from Table 6 (te Kamp, 1977). Begemann (1977) recommends, for the reference tip, $S_1 = 0.7$, irrespective of the type of driven pile.

Typical Dutch practice is to adopt a limit value of $q_s = 0.12\,MN/m^2$.

Table 6   Values of S_1 for use in Equation (6)

Type of pile	S_1 (reference tip, R*)
Timber	1.2
Parallel-sided concrete or steel:	
$\quad$ flat end	0.6
$\quad$ pointed end	1.1
Driven *cast-in-situ*†	1.6
Open steel tube and H-pile	0.7

*For f_s measured with a Dutch friction sleeve tip (M), the values of S_1 are halved.
† S_1 may be higher if the concrete is rammed as the casing tube is withdrawn.

9.1.3 Ultimate shaft resistance from cone resistance

Ultimate shaft resistance is given by:

$$Q_s = \sum_0^L q_s \pi d\Delta L = S_2 \sum_0^L q_c \pi d\Delta L \qquad (7)$$

Schmertmann (1978), suggests values of S_2 which 'express some Dutch practice', as shown in Table 7.

The values of S_2 in Table 7 are reasonably consistent with values of S_1 in Table 6 if it is assumed that $f_s = 0.01\,q_c$ (i.e. $R_f = 1\%$).

Again, a limit value of $0.12 MN/m^2$ is adopted.

Table 7   Values of S₂ for use in Equation (7)

Type of pile	S_2 (reference tip, R^*)
Timber	0.012
Precast concrete	0.012
Precast concrete, enlarged base†	0.018
Steel displacement	0.012
Open-ended steel tube	0.008

*Values of f_s are halved if f_2 is from the Dutch friction sleeve tip (M).
† Only applicable in a dense group of piles (otherwise reduce to $S_2 = 0.003$ where shaft is narrower than base)

Te Kamp (1977) recommends a value of $q_s = \dfrac{1}{300} q_c$ (i.e. $S_2 = 0.0033$) for open-ended steel tube piles driven into fine to medium sand as found in the southern part of the North Sea, recognising this as conservative and applying a factor of safety of 1.5. Beringen *et al.* (1979) present this value of q_s as more generally applicable. Both papers restrict q_s to $0.12\,\text{MN/m}^2$.

For piles driven into alluvial soils in the Glasgow area, Thorburn and MacVicar (1970) reported the use of $S_2 = 0.005$ for sands and $S_2 = 0.007$ for silts (using the Dutch mantle cone, M).

9.1.4   Shaft resistance – Nottingham's method

The method of estimating Q_s proposed by Nottingham (Nottingham, 1975; Schmertmann, 1978) uses the expression:

$$Q_s = k_s \left[\sum_{l-0}^{8d} \left(\frac{l}{8d} \right) f_s \pi d \Delta l + \sum_{8d}^{l} f_s \pi d \Delta L \right] \tag{8}$$

where k_s = correction factor in sands

l = depth to f_s value considered in the first summation

f_s = unit side resistance from friction sleeve.

Where there is uncertainty about f_s values, f_s may be replaced with $0.007 q_c$.

The first summation in Equation (8) is in the form of a correction for depth of embedment, applied over a depth of $8d$. Values of the correction factor, k_s, are given in Figure 42.

In Nottingham's method, with a continuously tapered or step-tapered pile, the pile length is divided into appropriate increments of constant-diameter length having the same total perimetral area, and the same procedure is used as for a parallel-sided pile. However, the k_s applicable to each constant-diameter length is determined by using the l/d ratio at the bottom of each such length. Also, at each step of a step-tapered pile or at each imaginary step of a continuously tapered pile, an 'additional side friction' is assumed, equal to the average value of q_c, over the length, multiplied by the horizontal area of the step and a factor S_3, as given in Table 8. If a step occurs within the upper length of $8d$, the additional side friction for that step is multiplied by $l/8d$.

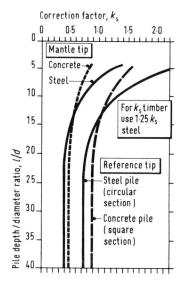

Figure 42 Nottingham's factor for calculating ultimate shaft resistance of a driven pile in sand (after Schmertmann, 1978)

Table 8   Nottingham's correction factor for additional friction along tapered and step-tapered piles

Soil	S_3	
	From Reference tip (R)	From Dutch friction sleeve tip (M)
Sand	1.6	1.6
Clay	1.0	0.6

9.1.5   Bearing capacity – the Poulos and Davis method

Poulos and Davis (1980) present a method of calculating the ultimate bearing capacity of a pile in sand which uses an idealised distribution of effective vertical stress adjacent to the pile, σ'_v, in which σ'_v is assumed to be equal to the effective overburden pressure at some critical depth, Z_c, beyond which σ'_v remains constant (see Figure 43).

Then
$$Q = \sum_0^L (F_\omega \pi d\sigma'_v K_s \tan \phi_a \Delta_z) + A_b \sigma'_{vb} N_q - W$$

where F_ω = correction factor for tapered pile (Figure 44) ($=1$ for uniform diameter)

K_s = coefficient of lateral pressure

N_q = bearing capacity factor (Figure 45), which is a function of the angle of shearing resistance of the soil below the base, ϕ_b

W = weight of pile

σ'_v = effective vertical stress along the shaft (limited to σ'_{vc} for $z > Z_c$)

σ'_{vb} = effective vertical stress at pile toe level

ϕ_a = angle of friction between pile and soil

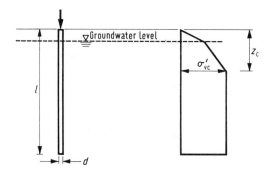

Figure 43 Simplified distribution of vertical stress adjacent to a single driven pile in sand (from Poulos and Davis, 1980)

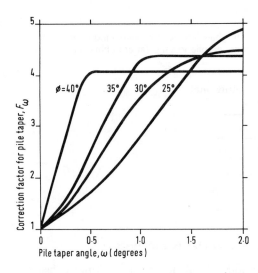

Figure 44 Pile taper factor related to pile taper angle (after Nordlund, 1963)

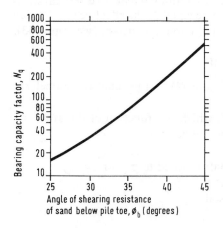

Figure 45 Bearing capacity factor plotted against angle of shearing resistance (after Berezantzev, Khristoforov and Golubkov, 1961). For driven piles, $\phi_b = \phi/2 + 20°$, for bored piles, $\phi_b = \phi - 3°$

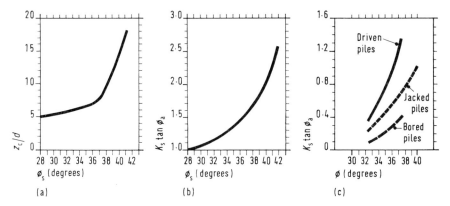

Figure 46 Shaft friction of piles in sand (after Poulos and Davis, 1980). For driven piles, $\phi_s=3\phi/4+10°$; for bored piles, $\phi_s=\phi-3°$

Values of Z_c/d and $K_s \tan \phi_a$ are plotted against values of ϕ_s in Figure 46(a) and (b). Values of $K_s \tan \phi$ are based on tests by Vesic (1967) on steel tube piles, and in their method Poulos and Davis make no distinction for other pile materials. The values of the angle of shearing resistance after installation, in the soil surrounding the shaft, to be used in Figure 46(a) and (b) are derived from values of ϕ prior to pile installation as given below. Values of ϕ can be determined from q_c as discussed in Section 5.3.

For driven piles, Z_c/d and $K_s \tan \phi_a$ are obtained from Figure 46 (a) and (b), N_q is obtained from Figure 45, and F_ω is obtained from Figure 44. Values of $K_s \tan \phi$ for driven piles proposed by Meyerhof (1976) are given in Figure 46(c). These values are considerably lower than the Poulos and Davis values in Figure 46(b). Values of $K_s \tan \phi$ for jacked piles are also given in Figure 46(c). The application of the Poulos and Davis method to bored piles is outlined in Section 9.2.

Poulos and Davis point out that their method may require modification in the light of future investigations. However, they applied it to 43 reported load tests on driven piles and found a mean ratio between calculated and measured ultimate loads of 0.98, with a standard deviation of 0.21. The ultimate load of all the piles considered was less than 3000kN, and they therefore suggest caution in considering piles of larger capacity.

9.1.6 Factors of safety and choice of method

The choice of factor of safety to be applied to ultimate pile bearing capacity depends on a number of factors, including reliability and sufficiency of site investigation data, confidence in the method of calculation, previous experience with piles in the same ground conditions, and whether or not pile-loading tests are to be undertaken. Where there are appreciable differences in CPT profiles, with a random spatial distribution, a reasonable lower bound should be adopted.

In cases where no specific estimation of settlement is to be made, the factor of safety may also be intended to limit settlements to reasonable values. Due allowance should then be made for the type of loading. This affects settlement (e.g. where repetitions of live loading are involved, settlement is greater than for single (or few) applications, particularly if the live load is large compared with the dead load).

Normal factors of safety as used in the Netherlands (te Kamp, 1977) for driven displacement piles directly designed from CPT results are given in Table 9.

Table 9   Normal factors of safety for driven piles in sand

Type of pile	Factor of safety (F)
Timber	1.7
Precast concrete:	
parallel sided	2.0
enlarged base	2.5

For his method, Nottingham recommends a factor of safety of 2.25 when using the reference tip (R) and 3.0 when using the Dutch friction sleeve tip (M). These are applied to both end bearing and shaft resistance.

The preferred method for calculation of ultimate end-bearing capacity, Q_b, was set out above in Section 9.1.1, taking account of the limiting values given. However, it is useful to check this against the Poulos and Davis method (see Section 9.1.5).

Where f_s values are capable of clear interpretation, it is recommended that ultimate shaft resistance, Q_s, should be calculated from both Equation (6), (page 71) and from Nottingham's method. Again, the Poulos and Davis method can provide a check. (If f_s is not capable of clear interpretation, Equation (7) (page 71) is used in place of Equation (6). The calculated values of Q_s are then each combined with the calculated value of Q_b, and the factors of safety appropriate to each method, as discussed above, are applied to arrive at values of safe bearing capacity:

$$Q_{safe} = \frac{1}{F} (Q_b + Q_s)$$

The lowest resulting value of Q_{safe} is then normally adopted. It may be necessary to modify the factor of safety in the light of the considerations in the first paragraph of Section 9.1.6.

The preceding discussion presupposes that pile loading tests are undertaken to confirm the pile design. However, on a small project, it may be considered uneconomic to carry out a pile-loading test prior to pile construction, in which case a higher factor of safety should be used.

Factors of safety for non-displacement piles are discussed in Section 9.2.

9.1.7   Shaft resistance of tension piles

The shaft resistance of a driven pile in sand is less when the pile acts in tension than when it acts in compression. Te Kamp (1977) suggests a unit ultimate shaft resistance $q_s = q_c/400$ in tension compared with $q_c/300$ in compression, a reduction factor of 0.75. Schmertmann (1978) proposes a reduction factor of 0.67, and for the special case of severely fluctuating tension loads he recommends that friction should be taken over only the middle half of the pile length, with a reduction factor of 0.33.

Begemann (1977) gives the equation:

$$Q_T = 0.7 \pi d \left[\sum_o^{L/4} f_s \Delta L + \tfrac{1}{3} \sum_{L/4}^{3L/4} f_s \Delta L + \sum_{3L/4}^{L} f_s \Delta L \right] \qquad (9)$$

where f_s is measured with a reference tip

Nottingham's method may also be used to calculate tension pile resistance, by multiplying the calculated ultimate shaft resistance in compression by 0.7.

Comparison of various methods of calculation with the results of three test piles (Beringen *et al.*, 1979) shows that Nottingham's method (with the reference test tip) gave results nearer to the observed test results than did the te Kamp method ($q_c/400$) or Begemann's method (Equation (9)). It is suggested that all three of these methods should be used and the lowest resulting value of ultimate tension resistance adopted, with a reasonably low factor of safety. Schmertmann's recommendation should be borne in mind for the special case of a severely fluctuating tension load.

9.2   Non-displacement piles in sand

Non-displacement piles include bored cast-in-situ piles, precast piles placed in a pre-bored hole, piles placed with the aid of jetting, and piles constructed by pumping sand-cement grout through the hollow stem of a continuous-flight auger. With bored piles, horizontal stresses decrease rather than increase as they do with a displacement pile. With pre-bored or jetted piles, any increase in stress is less than occurs with a driven parallel-sided pile, the reduction depending on the extent to which the pile is driven below the pre-bored or jetted depth. Stress increase may also be less with a pile which is vibrated into the ground or is cast within a vibrated open-ended casing. Hence non-displacement piles have lower shaft resistance than displacement piles of the same diameter. In bored piles, there may also be a reduction in end-bearing capacity because of loosening of soil below pile toe level.

Although some bored piles have been constructed in the Netherlands in recent years, they are few in number compared with driven piles. Hence there is no solid body of experience as there is with driven piles in sand. Pile capacities of non-displacement piles are sometimes calculated by the 'Dutch method', but a higher factor of safety (say 4) is applied. It is preferable to use the method of Poulos and Davis (Section 9.1.5), in which for bored piles, N_q and Z_c/d are obtained from Figures 45 and 46(a), using $\phi_s = \phi_b = \phi - 3°$ to allow for possible loosening of the surrounding and underlying soil during installation. $K_s \tan \phi_a$ is obtained from Figure 46(c). For bored piles larger than 0.5m in diameter, the Meyerhof reduction factor, R_b, should be applied to the calculated ultimate end-bearing capacity:

$$R_b = \left(\frac{d+0.5}{2d}\right)^n$$

where n = 1 for loose sand

        n = 2 for medium dense sand

and   n = 3 for dense sand

For $d > 1.5$m, R_b is taken as for $d = 1.5$m

No specified value of factor of safety is proposed by Poulos and Davis. It is therefore suggested that a value of 2.5 should be used.

Because of the uncertainties concerning non-displacement piles in sand, and the considerable effect which installation procedures can have on bearing capacity and settlement, it is recommended that pre-construction pile-loading tests should be undertaken. It might be feasible to dispense with these on very small projects where there is considerable local experience, but in such cases factors of safety should be increased.

9.3   Settlement of piles in sand

At present, there is no direct method of calculating the settlement of a pile from CPT data. However, there are some indirect methods.

An approximate estimate of the settlement of a single pile in sand can be obtained from Meyerhof's (1959) equation:

$$s_1 \simeq \frac{d_b}{30F}$$

where d_b = diameter of pile base

and   F = factor of safety on ultimate load (> 3)

This may be sufficient in many cases, but if a more detailed analysis is required it is necessary to determine values of Young's modulus, E, and Poisson's ratio, v. Poisson's ratio for sands is usually between 0.25 and 0.35, and a value of 0.30 can be adopted without significant error. E can be evaluated from q_c as described in Section 5.4.2. However, driving a pile into loose or medium dense sand gives rise to local increase in relative density and Young's modulus, so that values of E relevant to the settlement of a single pile may be higher than would be determined from Figure 17 (Section 5.4). Hence the resulting estimate of settlement is conservative.

Description of the methods of analysis is outside the scope of this Report, and reference should be made elsewhere (e.g. Poulos and Davis, 1980). These authors also describe methods of estimating settlements of pile groups in sand, for which a value of E is again required. Another useful reference is Vesic (1977).

Empirical relationships between settlement of a group of driven piles in sand and the settlement of a single pile are given by Skempton (1953) and by Meyerhof (1959).

9.4 Piles in clay

Although methods are available for calculating the bearing capacity of piles in clay in terms of effective stress parameters (e.g. Burland, 1973), it is more usual to use the undrained shear strength, c_u. This can be obtained from CPT results using the methods described in Sections 6.2 and 6.3. At present, there are no commonly adopted procedures for determining pile bearing capacity in clay direct from CPT results, and other methods are preferred.

If CPT results are used to obtain values of c_u for calculation of the bearing capacity of piles in OC clay, it should be remembered that the values derived by the methods of 6.3 relate to shear strength back calculated from plate-loading tests. For end-bearing calculations, it is suggested that the derived values are directly used, without the customary reduction in c_u made to allow for overestimation of c_u measured on small laboratory specimens. However, for calculation of shaft resistance, where empirical values of the adhesion factor are based on undrained shear strength determined on small laboratory speciments, $c_{u\ (PLT)}$ values derived from q_c have to be converted to laboratory values. For London Clay:

$$c_{u(lab)} \simeq (1 + 0.04Z) . c_{u(PLT)}$$

where the relationship between laboratory c_u and $c_{u\ (PLT)}$ is not known, $c_{u\ (PLT)}$ can be used as a conservative value of c_u.

9.5 Piles in calcareous soils

In respect of Bombay calcareous soils, Beringen *et al.* (1982) concluded that, for the calculation of pile-bearing capacity, conventional methods using CPT are appropriate. Nevertheless, it is imprudent to assume that this applies to other calcareous soils, particularly those with a higher carbonate content than the Bombay soils. There is some experience of the load-bearing capacity of piles in calcareous soils being considerably lower than would be expected. At this stage, it is recommended that the design of piles in carbonate soils should be based on pile-loading tests.

9.6   Synopsis: Piles

Refer to Section 9.1.

9.6.1   Driven piles in sand

For choice of method and factor of safety, refer to Section 9.1.6.

(a) *Ultimate end bearing*

Refer to Section 9.1.1.

Dutch method:

$$Q_b = q_p A_b$$

where $q_p = \dfrac{q_{c1} + q_{c2}}{2}$ (see Figure 40)

A reduction in Q_b may be necessary if there are weaker layers between 4 and 10 pile diameters below the pile toe level.

Limit values of q_p are shown in Figure 41.

For usual factors of safety, see Table 9. Limit value of q_c is usually 30 MN/m².

(b) *Ultimate shaft resistance*

Refer to Sections 9.1.2 and 9.1.3.

(i) Dutch method:

$$Q_s = \sum_0^L q_s \pi d \Delta L = S_1 \sum_0^L f_s \pi d \Delta L$$

q_s usually limited to no more than 0.12 MN/m².

then
either use local side friction, f_s, in:

$$Q_s = S_1 \sum_0^L f_s \pi d \Delta L$$

Refer to Table 6 for values of S_1 for different types of pile or cone.

or,
if f_s difficult to interpret, use cone resistance, q_c, in:

$$Q_s = S_2 \sum_0^L q_c \pi d \Delta L$$

Refer to Table 7 for values of S_2 for different types of pile or cone.

(ii) Nottingham's method:

Refer to Section 9.1.4.

For parallel-sided piles

Refer to text and Table 8 for tapered and step-tapered piles.

$$Q_s = k_s \left[\sum_{l=0}^{8d} \left(\frac{l}{8d} \right) f_s \pi d \Delta L + \sum_{8d}^{l} f_s \pi d \Delta L \right]$$

Refer to Figure 42 for values of k_s for different types of pile or cone.

Recommended factors of safety are:
$F = 2.5$ for use with reference tip (R) results,
$F = 3.0$ for use with Dutch friction sleeve cone (M) results, but see also Section 9.1.6.

(c) *Bearing capacity by Poulos and Davis method*

Refer to Section 9.1.5.

Evaluate Z_c from Figure 46(a) and estimate ϕ (see Section 5.3)

$$Q = \sum_0^L (F_\omega \pi d \sigma_v' K_s \tan \phi_c \Delta z) + A_b \sigma_{vb}' N_q - W$$

Caution needed if $Q > 3000$ kN

where σ_v' is effective vertical stress along shaft and $K_s \tan \phi_a$ is derived from Figure 46(b) using ϕ_s
$F_\omega = 1$ for uniform pile section

σ_v' limited to σ_{vc}' for $z > z_c$ (Figure 43)
$\phi_s = \frac{3}{4}\phi + 10°$
Refer to Figure 44 for values of F_ω for tapered piles.

N_q derived from Figure 45, using ϕ_b

$\phi_b = \dfrac{\phi}{2} + 20°$

(d) *Tension piles*

Adopt lowest result from the methods below with a reasonably low factor of safety.

Te Kamp's method:
$q_s = q_c/400$

Begemann's (1977) method:

$$Q_t = 0.7\pi d \left[\sum_0^{L/4} f_s \Delta L + \tfrac{1}{3}\sum_{L/4}^{3L/4} f_s \Delta L + \sum_{3L/4}^{L} f_s \Delta L \right]$$

Nottingham's method
$\phi_s = 0.7\,(Q_{s(compression)})$

Refer to Section 9.1.7.

f_s measured with reference tip.

For severely fluctuating tension, see page 77.

9.6.2 Non-displacement piles in sand

Use Paulos and Davis method

$\dfrac{Z_c}{d}$ derived from Figure 46(a) using ϕ_s

N_q derived from Figure 45 using ϕ_b
$K_s \tan \phi_a$ derived from Figure 46(c), using $\phi = \phi_s$
For $0.5 < d < 1.5$ m apply reduction factor, R_b, to ultimate end bearing capacity.

$$R_b = \left(\frac{d+0.5}{2d} \right)^n$$

For $d > 1.5$ m, R_b as for $d = 1.5$ m.
In combination with pre-construction pile loading tests, a factor of safety of $F = 2.5$ should be used: without these loading tests, F should be higher.

Refer to Section 9.2.

$\phi_s = \phi_b = \phi - 3°$

$n = 1$ for loose sand.
$n = 2$ for medium dense sand.
$n = 3$ for dense sand.

9.6.3 Settlement of piles in sand

An approximate estimate of settlement of a single pile is

$$s_1 \simeq \frac{d_b}{30F}$$

for $F > 3$

Detailed analysis requires determination of E and v (may be taken as 0.3). For evaluation of E, see Section 5.4.2.

For settlement of pile groups, refer to empirical relationships of Skempton (1953), Meyerhof (1959), and analysis method of Poulos and Davis (1980).

Refer to Section 9.3.

No direct method of calculating pile settlement from CPT data.

Refer to Poulos and Davis (1980) and Vesic (1977) for methods of analysis.

Driving a pile into loose or medium dense sand increases E, hence calculated settlement maybe conservative.

9.6.4 Piles in clay

There are no commonly adopted procedures based on CPT results for piles in clay.

Refer to Section 9.4.

9.6.5 Piles in calcareous soils

Design should be based on pile-loading tests.

Refer to Section 9.5

10 Other applications

10.1 Control of compaction and ground improvement

10.1.1 General

The main use of the CPT in relation to compaction and ground improvement is in the control of deep compaction of mainly granular soils. For shallow compaction, various other control methods are preferred, although the CPT can be useful in checking the variability of a fill compacted in layers, or in checking whether unsatisfactory material has been left below a fill. In the improvement of cohesive soils by means of surcharge, with or without vertical drains, the prime task is the monitoring of the rate of dissipation of excess pore pressures, and although the cone penetrometer and piezocone have a role in design, piezometers are the principal tool for monitoring. Changes in undrained shear strength in cohesive soils are frequently measured by *in-situ* vane testing, but the CPT may also be used because of its rapidity and continuity of measurement. A good combination for such work is a number of vane tests, and laboratory tests on samples from boreholes, with CPTs to interpolate between them.

Many ground improvement processes are carried out on a grid pattern. Examples are vibro-flotation and vertical drains placed below a surcharge fill. In these cases, the degree of improvement achieved may vary with distance from the grid points, and it is therefore important that any compliance testing by CPT should take this into account.

10.1.2 Deep compaction of granular soils

Compaction of granular soils in depth may be achieved by vibro-flotation, dynamic compaction (heavy tamping), insertion and backfilling of tubes by driving or top-vibration, driving of piles, compaction grouting, blasting or other special measures. Control methods are similar for all these techniques, and they include SPTs in boreholes, dynamic sounding, CPT, and pressuremeter. In the control of dynamic compaction, over-all settlement of the ground surface is monitored, and the effect of the compactive effort is measured before, sometimes at various stages during, and after compaction. In this connection, it is often desirable to repeat measurements subsequent to completion of compaction, because improvement may continue to occur for some weeks.

The pressuremeter has frequently been used to monitor dynamic compaction, but it is relatively slow in operation, and it involves some uncertainties arising from disturbance during installation, or from the use of a split metal sheath during installation and testing where the ground conditions require it. Thus, in recent years, the CPT has come into greater use either in conjunction with or replacing the pressuremeter. Nevertheless, it is desirable that more than one *in-situ* testing technique should be adopted on a project, at least in the early stages (e.g. CPT and SPT, or CPT and pressuremeter). Plate-loading tests or area-loading tests may also be required to confirm estimates of settlement based on CPT or other test results. Figure 47 shows an example of cone resistance, SPT, and pressuremeter limit pressure measurements before and after compaction. The ground consisted of some 2 m of fill overlying sabkha, sandy silt in the upper part and silty sand below.

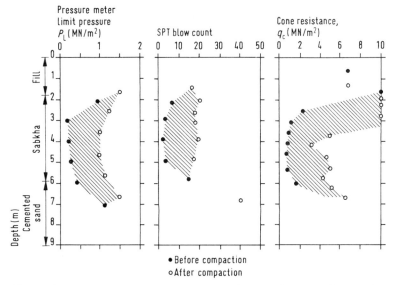

Figure 47 Example of cone resistance, standard penetration test values, and pressuremeter limit pressure before and after dynamic compaction (from Juillie and Sherwood, 1983)

Compliance criteria for deep compaction are sometimes stated in terms of relative density, particularly where compaction is required to improve resistance to liquefaction during earthquake shaking. However, it should be borne in mind that relative density in this context is relative density as assessed by the SPT, because almost all experience of liquefaction is related to the SPT. Moreover, this is the SPT as measured with the slip-rope method in a small diameter borehole. This should not be confuse with relative density as determined by comparing *in-situ* density with values of maximum and minimum density measured in the laboratory, or with the results of *in-situ* tests which purport to measure relative density in these terms (nuclear density meters, conductivity probes, etc.).

If liquefaction is the problem, the criteria for ground improvement should be in terms of the SPT or its equivalent CPT values (see Section 5.5). Similarly, when the problem is one of settlement under static loading, the criteria should be related to deformation parameters to be derived from the *in-situ* control testing methods employed and not to a concept of relative density.

10.2   Liquefaction potential

The liquefaction of a saturated sand deposit subjected to earthquake shaking can be assessed either from an analysis of stress or strain conditions combined with laboratory testing, or empirical correlations between observed behaviour and some parameter measured *in situ*. The procedure based on laboratory testing suffers the disadvantage that it is extremely difficult (if not impossible) to obtain truly undisturbed samples, and the test results are very sensitive to sample disturbance. For *in-situ* testing, the CPT has advantages over the SPT in that it is not so sensitive to variations in operational procedures and that it provides a continuous record. The CPT can also serve as an aid to identification, which is important in assessing liquefaction potential. On the other hand, there is a large body of data relating observed performance during earthquake shaking to SPT results, but very little data relating it to CPT results. The main basis for evaluation of liquefaction potential using CPT results is therefore through converting them to equivalent SPT values.

10.2.1   Liquefaction potential via SPT

A basis for conversion of CPT values to SPT values is given in Figure 19 (Section 5.5). However, there are uncertainties involved.

The method of assessment of liquefaction potential from SPT values is outlined by Seed and Idriss (1971, 1981). See also Robertson and Campanella (1985).

Davis and Berrill (1983) suggest that Seed's method overestimates the effect of small earthquakes and underestimates the effect of large earthquakes.

10.2.2   Liquefaction potential direct from CPT

A start has been made in obtaining direct correlations between CPT data and observed performance by studies made in China at sites affected by the Tangshan earthquake, which had a magnitude of 7.5 on the Richter Scale (Zhou, 1980, 1981). This led to a method in which a critical value of cone resistance, q_{crit}, is given by:

$$q_{crit} = q_{co}[1 - 0.065 \ (H_w - 2)][1 - 0.05(H_o - 2)]$$

where H_o = thickness of cohesive overburden layer (non-liquefiable) (m)
         H_w = depth to groundwater level (m)
and q_{co} is a function of shaking intensity as given below:

Modified Mercalli intensity	VII	VIII	IX
Maximum surface acceleration (Chinese code)	0.1g	0.2g	0.4g
q_{co}(MN/m^2)	4.6	11.5	17.7

The penetrometer tip used in arriving at the above correlation had a diameter of 45mm, with a reduced diameter above the friction sleeve, so that it gave cone resistances somewhat different from a reference tip. The soil involved ranged from medium to coarse sand to silty (or very silty) fine sand, with D_{50} in the range 0.55 to 0.07mm. For soils with a fines content greater than about 30%, the correlation does not apply (Zhou, 1981).

Part 2

The piezocone

11 Piezocones and their use

11.1 Historical outline

Measurement of the porewater pressures developed by advancing a probe into the ground and of their dissipation were first made in Sweden in the early 1970s (Torstensson, 1977). The probes were on the lines shown in Figure 48. No measurement of cone resistance and skin friction was possible. In 1973, the Norwegian Geotechnical Institute introduced a pore-pressure probe of the same shape as the reference tip, but again only pore-pressure measurement could be made, and it was

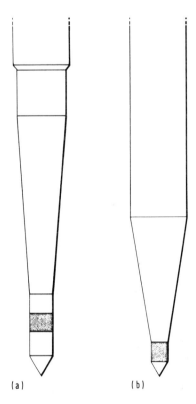

(a)   (b)

Figure 48 Pore-pressure probes (a) From Torstensson, 1977. (b) From Wizza *et al.*, 1975

necessary to carry out a separate CPT in order to combine the measurements of pore pressure and cone resistance (Janbu and Senneset, 1974). Alternatively, the pore-pressure results were correlated with adjacent borehole data.

In a separate development in 1969, a piezometer element was incorporated into a 50-kN capacity cone penetrometer, without friction sleeve, for the purpose of measuring the excess pore pressures in sand layers below the Dutch polders (Zuidberg *et al.*, 1982). The porous element was placed in the shaft of the tip, immediately above the base of the cone. Subsequent developments involved the combination of pore pressure measurement with measurement of both cone resistance and skin friction.

11.2   Role of the piezocone

Applications of the piezocone fall into two groups: profiling and assessment of engineering parameters. Profiling with the piezocone gives more accuracy than the CPT, but as yet it has not led to any major improvement in identification of soils.

Applications in assessment of engineering parameters include:

1. assistance in the interpretation of cone resistance and skin friction in terms of shear strength and deformation characteristics
2. assessment of *in-situ* permeability and consolidation characteristics
3. assistance in the assessment of stress history and OCR of cohesive soils
4. measurement of static porewater pressures.

The last of these applications is relatively straight forward, but the other three are in the early stages of development, without a supporting body of experience. Their inclusion here is in recognition of their potential advantages. Hence any of the associated procedures outlined in Part 2 of this Report should be used with caution. Wherever possible, the results should be checked against alternative investigation methods.

Piezocone sounding is particularly useful in mixed deposits where it is often difficult with the CPT to know whether the data refer to drained or undrained conditions.

11.3   Equipment of the piezocone (CPTU)

The pore-pressure probe has been largely superseded by the piezocone, incorporating a porous element into the tip. There is no standardisation at present. Different designs are in use, the principal difference being the location of the porous filter element (see Figure 49). The ISSMFE Technical Committee is considering the inclusion of the piezocone in the IRTP to be published in 1989. Whether or not there will be sufficient support for this is uncertain, but it is expected that at least the geometry of the reference tip will be specified in a way which will allow the presence of a porous element to be included in any of the positions shown in Figure 49. The arguments

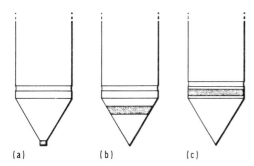

(a)   (b)   (c)

Figure 49 Positions of porous elements on piezocones. (a) At cone point. (b) On face of cone. (c) On shaft, between cone and sleeve

concerning filter location involve both practical issues and consideration of the magnitude, variation and causation of the pore pressure developed.

The magnitude of the measured pore pressure varies considerably with the position of the filter, as illustrated in Figure 50. It can be seen that the maximum pore pressure is developed on the cone face without much difference in value between the point of the cone and higher up the face, and that there is a rapid fall off in pore pressure with distance above the base of the cone.

Clogging and excessive wear appear to be serious problems with filters located at the point of the cone. A filter located at or close to the mid height of the cone face has the advantage that it is in a zone of maximum and relatively constant pore pressure. Furthermore, pore-pressure dissipation is probably more rapid from this position than from above the cone (Acar *et al.*, 1982).

A filter located well up the shaft is also in a position of stable pore-pressure response. It gives somewhat similar results to a filter on the cone face, but with a reduced amplitude in the peaks and troughs of excess pore pressure. Wroth (1984) points out that pore pressure measured on the shaft has a maximised component of pore pressure because of shear, whereas that measured on the face includes a large component resulting from the increase in octahedral stress. Thus, measurement high on the shaft has advantages in respect of determination of the stress history of a deposit, because, during shear, NC cohesive soils tend to develop large positive pore pressures, whereas OC cohesive soils develop smaller (even negative) pore pressures. However, at this location, the pore-pressure response is largely controlled by severely remoulded soil, and such response may not therefore be relevant. Furthermore, the pore-pressure response varies with stiffness ratio (see Section 6.1) as well as OCR.

There are practical difficulties involved with the positioning of the porous element on the face of the cone. Jones and van Zyl (1981) report clogging of filters in clay when using four circular filters on the face of the cone. However, Tümay *et al.* (1981) found disc-shaped aluminium oxide filters at mid height of the cone to be satisfactory. Ceramic filters appear to be less prone to clogging than stainless steel filters, which are more resistant to wear. A detailed discussion of the factors to be considered in the selection of a filter element is given by Jamiolkowski *et al.*, 1985.

UK practice is both limited and varied at present. Some organisations use disposable ceramic filters at mid height of the cone face. The filter wears during penetration, thus reducing clogging (Figure 51). Others use different types, including

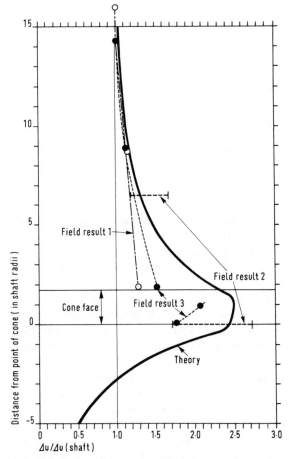

Figure 50 Distribution of penetration pore pressure (data from Campanella, Gillespie and Robertson, 1982; Levadoux and Baligh, 1980; Sugawara and Chikaraishi, 1982)

a tip with a porous element above the cone. The ideal piezocone appears to be one in which pore pressures are measured both on the face or shoulder of the tip and on the shaft. Tips with multiple piezometric elements on the face of the cone and on the shaft are now commercially available.

The frequency of changes in pore pressures is such that the transducer should have a minimum time lag. Volume change of the transducer should not exceed $2.5 \, \text{mm}^3$ per kN/m^2 (Silver, 1979). The gap between the porous element and the transducer should be kept to a minimum. Where cone resistance is more than about $5 \, \text{MN/m}^2$, pore-pressure measurement may be affected by the compressibility of the porous element, and of the cone (Jamiolkowski *et al.*, 1985).

When measuring pore-pressure dissipation, it is considered necessary to clamp the rods in the penetrometer rig to avoid stress relief at the cone face, which leads to a drop in pore pressure. This does not apply in the case of the filter placed in the shaft of

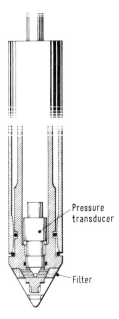

Pressure
transducer

Filter

Figure 51 Example of piezocone in use in the United Kingdom (after Swain, 1984)

the penetrometer tip. Pore pressures vary considerably with rate of penetration, and it is important that the standard rate of 20 mm/s should be closely adhered to. Frequent calibration of the pore-pressure transducer should be made.

Experience shows that some soils (e.g. stoney glacial tills) can produce negative porewater pressure readings from piezocones, particularly for elements just above the cone. In this case, it is necessary to use either silicone oil or glycerine as the liquid between the porous element and the transducer, because, if water is used, the negative pore pressure sucks the water out and leaves an air pocket, thus impairing pore-pressure measurement for the rest of the profile.

11.4 De-airing

It is essential that the pore-pressure measuring system should be saturated, because even a very small quantity of entrained air significantly increases the compressibility of the fluid and seriously affects the response time of the system. At present, problems with de-airing are the main obstacle to more extensive use of the piezocone. Various ways have been adopted to ensure saturation, including:

1. de-airing of the system under vacuum for a period of 12 h or more, and the use of de-aired water, together with some means of retaining it within the filter until the tip is below water level (e.g. enclosing the tip in a thin plastic membrane which shears off when entering the soil). This may require pre-boring down to groundwater level.
2. use of glycerine as the system fluid (Campanella *et al.*, 1981, Jones *et al.*, 1981) or

de-aired silicone oil (Battaglio *et al.*, 1981, Smits, 1982). Alcohol has been used in cold climates (Wizza *et al.*, 1975).

Lack of success in de-airing may be indicated by an unnaturally smooth pore-pressure profile. The profile should only be smooth in homogeneous NC clays. It may also be indicated by an unusually long period between a halt in penetration and the re-establishment of equilibrium pore pressure (say 200 min or more, instead of less than 50 min for a fully de-aired tip).

11.5   Profiling and identification with the piezocone

The pore-pressure response of the piezocone is such that thin layers can be identified. Where there is a good contrast in permeability of adjacent layers, the thinnest

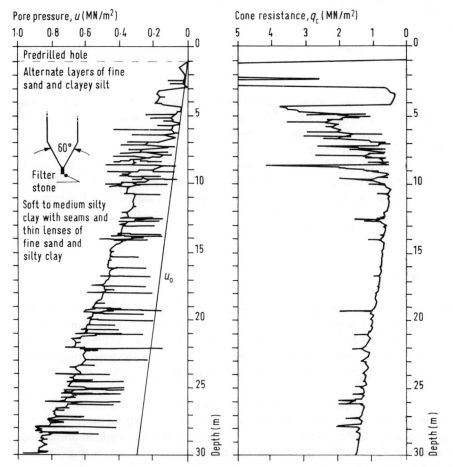

Figure 52 Example of piezocone measurements in a cohesive formation with a highly developed macrofabric, Porto Tolle (Italy) (after Battaglio and Maniscalco, 1983)

identifiable layer is 30 to 50 mm thick. An example of the results of a CPT with pore-pressure measurement is given in Figure 52 (Battaglio and Maniscalco, 1983).

Further examples of pore-pressure profiles (using a tip with the porous element on the shaft, immediately above the cone as in Figure 49(c)) are given in Figure 53(a) and (b). The soils concerned were probably of marine origin, with OCR between 1 and 4, typically between 2 and 2.5. Figure 53(a) shows the results of four profiles, obtained along a proposed road axis, in a clay of extreme homogeneity in both the vertical and horizontal directions, with constant values of the correlation factors $N_k = (q_c - \sigma_{vo}/c_u$ and $N_{\Delta u} = \Delta_{u/cu}$. Figure 53(b) shows profiles in two clay strata separated by a sand sequence. In the upper clay stratum, there is strength variation horizontally, and sand lenses are present.

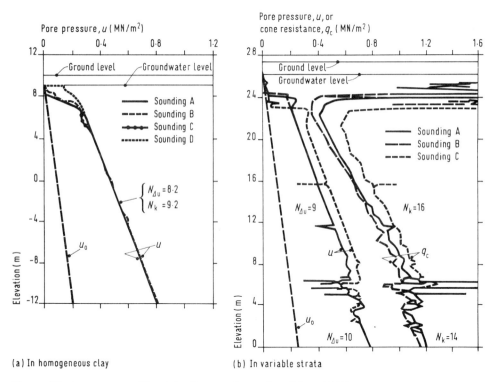

(a) In homogeneous clay   (b) In variable strata

Figure 53 Piezocone profiles (after Tavenas, Leroueil and Roy, 1982)

The piezocone is also useful in the investigation of landslides (Torstensson, 1979). Partly remoulded slide debris reduces values of q_c and Δ_u, and failure planes are shown by low values of excess pore pressure.

Unfortunately, it is not yet possible to produce a positive correlation between pore-pressure measurements during a CPT and soil type. A tentative correlation based on the change in pore pressure, Δ_u, and the net cone resistance, $(q_c - \sigma_{vo})$ is presented by Jones and Rust (1982) for use with a penetrometer tip with the filter placed immediately above the base of the cone (Figure 54).

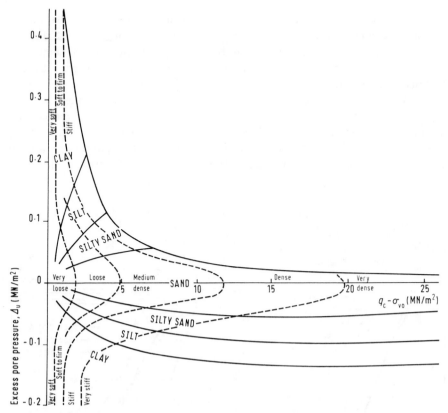

Figure 54 Tentative soil type identification by piezocone (porous element placed on shoulder immediately above base of cone) (after Jones and Rust, 1982)

A broad classification (derived from only limited data) proposed by Senneset and Janbu (1985), based on their dimensionless parameters, N_m and B_q, as defined in Section 12.2.3, is presented in Figure 55. Additional help is given by Table 10, which is based on their parameter, a, the 'attraction', and $\tan \phi'$, as expressed in:

$$\tau_f = (\sigma' + a) \tan \phi'$$

were τ_f is the shear stress at failure.

The Senneset and Janbu classification relates only to a filter placed immediately above the base of the cone, and it is not applicable to a cone with a filter in a different position. Results from a recently completed comparative study of different types of piezocones in stiff OC clays (BRE/NGI, 1985) show that while positive pore pressures were generated on the face of the cone, negative pore pressures were recorded by cones with filter elements behind the cone. Thus B_q values calculated from pore pressure measured behind the cone may be zero to negative. This differs from the Senneset and Janbu correlation which indicates positive values of B_q. The cones used in the comparative study had sintered stainless steel filters and used either de-aired water or silicone oil as a saturation fluid.

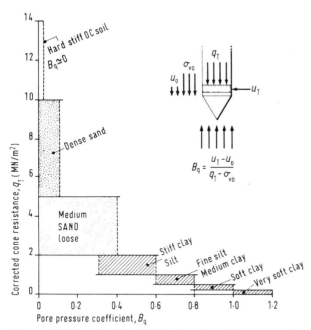

Figure 55 Tentative classification chart based on cone resistance and pore-pressure ratio (after Senneset and Janbu, 1985)

Table 10   Typical values of attraction and tangent of effective angle of shearing resistance (after Senneset and Janbu, 1985)

Soil state	Clay		Silt		Sand	
	a (kN/m²)	$\tan \phi'$	a (kN/m²)	$\tan \phi'$	a (kN/m²)	$\tan \phi'$
Soft–loose	5 to 10	0.35 to 0.45	0 to 5	0.50 to 0.60	0	0.55 to 0.65
Medium	10 to 20	0.40 to 0.55	5 to 15	0.55 to 0.65	10 to 20	0.60 to 0.75
Stiff, dense	20 to 50	0.50 to 0.60	15 to 30	0.60 to 0.70	20 to 50	0.70 to 0.90

Using a cone with a filter on the face (see Figure 51), Fugro Ltd. found that, in clays, B_q appears to depend on OCR rather than on shear strength. For example, in a 22-m thick layer of NC or lightly OC North Sea clay (British sector, northern part), consistency varied between very soft at the surface to firm below a depth of about 13 m, whereas B_q was virtually constant with depth at a value of about 1.1.

11.6   Equilibrium pore pressures

Profiles of groundwater pressure frequently differ from the simple case of a hydro-static distribution below a well-defined phreatic surface. Under-drainage may occur below a clay stratum, or there may be a series of perched watertables within a generally permeable profile. The pore-pressure probe of the piezocone can be used to elucidate the piezometric profile by allowing, at given stages, the pore pressure developed to dissipate to an equilibrium or near equilibrium value. This may take considerable time in a cohesive soil, and it may be preferable to insert a number of static piezometers at the start of a programme of piezocone testing.

12 Soil parameters from the piezocone

12.1 General

As has already been pointed out, the use of the piezocone for the assessment of engineering parameters is in an early stage of development. The discussion which follows illustrates how CPTU results might be useful, and reflects the current stage of development.

It is important that the supervision and interpretation of CPTU investigations should be undertaken by an experienced geotechnical engineer.

12.1.1 Correction of CPT values

It has recently been realised that, in theory, it is necessary to correct cone resistance and skin friction for pore-pressure effects. Water pressure can act over an area at the base of the cone and on the ends of the friction sleeve. The effect on cone resistance is shown in Figure 56.

The extent of the correction varies with the detailed design of the tip.

$$q_T = q_c \text{ (corrected)} = q_c \text{ (measured)} + \lambda u_T \tag{10}$$

where u_T = total pore pressure at the base of the cone
$\lambda = A_g/A_c$
A_g = area of groove
and A_c = projected base area of the cone

For the reference tip geometry, λ is generally in the range from 0.15 to 0.25. Similarly, a correction is applicable to the friction sleeve readings:

$$f_s \text{ (corrected)} = f_s \text{ (measured)} + \frac{u_{sh} \cdot A_{sh} - u_T \cdot A_t}{A_s}$$

Although, in theory, such corrections are necessary, in practice they are at present seldom used in published data. The correction is relatively insignificant in granular soils, but it can be important in cohesive soils, particularly in soft clays. The effect of high water pressure in overwater soundings is limited by setting readings to zero at the start of penetration.

It is recommended that the λ value should be quoted on all result sheets.

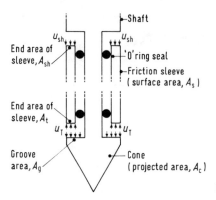

Figure 56 Correction of penetrometer readings

12.1.2   Pore-pressure ratio

Various definitions of pore-pressure ratio have been proposed, including:

u_T/q_c	(Baligh *et al.*, 1981)
Δ_u/q_T	(Campanella and Robertson, 1981)
$\dfrac{\Delta_u}{q_c - u_o}$	(Smits, 1982)
$\dfrac{\Delta_u}{q_c - \sigma_{vo}}$	(Jefferies and Funegard, 1983)
$\dfrac{u_T - u_o}{q_T - \sigma_{vo}}$	(Senneset and Janbu, 1985)

where u_T = total pore pressure, $\Delta_u = u_T - u_o$, and q_T = corrected cone resistance.

It seems likely that the last definition may eventually be adopted as a standard.

12.2   Shear strength

12.2.1   Undrained shear strength

The determination of undrained shear strength from pore-pressure soundings is complicated by the fact that Δ_u/q_c appears to vary with OCR (see Section 12.2.2), so that, at a given shear strength but at different values of OCR, a clay gives different values of Δ_u/q_c. Also, as already discussed, the pore pressures measured during penetration vary with the position of the filter, which complicates comparison of results. Considerable caution is therefore required when attempting determination of shear strength from pore-pressure soundings.

Relationships between the generated *excess* pore pressure during cone penetration and undrained shear strength, are proposed by various authors. Theoretical or semi-theoretical approaches using cavity expansion theory are considered by Vesic (1972), Battaglio *et al.* (1981), Randolph and Wroth (1979), Henkel and Wade (1966), and others.

Empirical relationships between Δ_u and c_u are usually presented in the form:

$$\Delta_u = N_{\Delta u} \cdot c_u$$

Using the spherical cavity expansion theory (Vesic, 1972), $N_{\Delta u}$ is in the range from 4 to 7, and from cylindrical cavity expansion theory, in the range from 3 to 5.

For pore pressure measured at the base of the cone, and c_u by *in-situ* vane, Tavenas *et al.* (1982) give as a typical correlation, based on investigation of some Canadian clays:

$N_{\Delta u} = 7.9 \pm 0.7$, for $0.8 < I_L < 2.0$
and $N_{\Delta u} = 11.7 \pm 2.0$, for $I_L > 2.0$
where I_L is liquidity index

However, the authors point out that the correlations are too scattered to permit an accurate evaluation of c_u for design purposes.

For Δ_u measured at the cone point and c_u by *in-situ* vane, for St. Albans' Clay (a soft sensitive marine clay with an OCR of 2.1 to 2.3), Roy *et al.* (1982) found $N_{\Delta u} = 7.4$ approx. For Δ_u measured one diameter above the base of the cone, $N_{\Delta u}$ fell to 4.7.

Somewhat lower values of $N_{\Delta u}$ are to be expected in firm or stiff clays of low sensitivity.

At this stage, local correlations are to be preferred.

12.2.2 Overconsolidation ratio

Roy *et al.* (1981) suggest that Δ_u may be directly related to preconsolidation pressure σ'_p, and hence to OCR. For St. Albans' Clay, and for Δ_u measured at the point of the cone, they found $\Delta_u = 1.72\sigma'_p$ and for Δ_u measured on the shaft, $\Delta_u = 0.93\sigma'_p$.

However, a comparison by May (1983) of various proposed relationships (Baligh *et al.* 1981, Lacasse and Lunne, 1982, Lacasse *et al.* 1981, Smits, 1982 and Tümay *et al.*, 1981), as shown in Figure 57(a) suggests that different clays exhibit markedly different behaviour, probably because the ratio of u or Δ_u to q_c is a function of G/c_u where G is the shear modulus of deformation, rather than of OCR.

A relationship between pore-pressure ratio, using pore pressure and net corrected cone pressure, and OCR is shown in Figure 57(b) (Coutts, 1986). This derives from tests with a filter on the face of the cone, at six sites, five over water and one on land.

12.2.3 Effective stress shear strength

At present, there is no generally applicable, reliable method of determining effective stress shear strength parameters of cohesive soils from CPTU data. However, the theoretically derived method by Senneset and Janbu (1985) appears to give reasonable effective stress shear strength parameters in soft NC clays. In their method, effective stress shear strength is expressed in the form:

$$\tau_f = (\sigma' + a) \tan \phi'$$

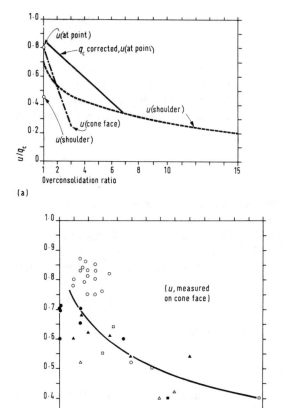

(a)

(b)

Figure 57 Variation of pore-pressure ratio with overconsolidation ratio (a) Using total pore pressure and uncorrected cone resistance (after May, 1983). (b) Using excess pore pressure and net corrected cone resistance (from Coutts, 1986)

where a, the 'attraction', is the negative intercept on a plot of net cone resistance against effective overburden pressure, as shown in Figure 58(c). Two dimensionless parameters are used:

cone resistance number, $N_m = \dfrac{q_T - \sigma_{vo}}{\sigma'_{vo} + a} = \dfrac{q_n}{\sigma'_{vo} + a}$

pore-pressure ratio, $B_q = \dfrac{u_T - u_o}{q_T - \sigma_{vo}} = \dfrac{\Delta_u}{q_n}$

Tan ϕ' is then read from Figure 58(d).

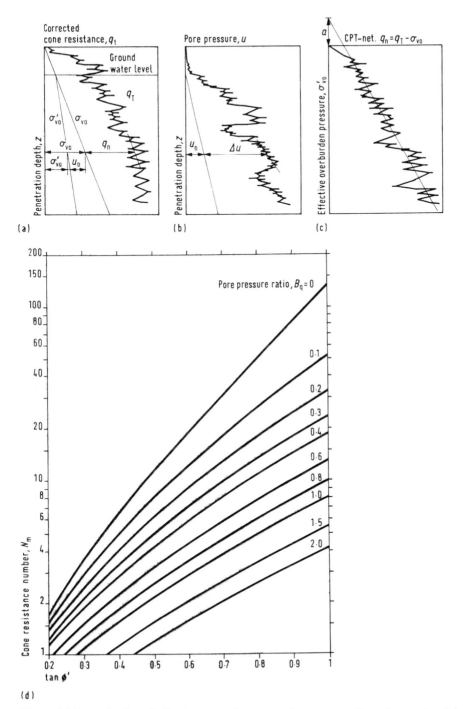

Figure 58 Determination of effective stress shear strength parameters (from Senneset and Janbu, 1984)

A method of determining ϕ' of NC mine tailings is proposed by Sugawara and Chikaraishi (1982). They use a penetrometer tip, having a recess in the shaft situated 50mm above the base of the cone, with a pore-pressure filter immediately above the cone base. They also distinguish average grain size of the tailings on the basis of pore-pressure ratio, Δ_u/σ'_v.

Although this method is expressly for use in mine tailings, it may be useful in similar NC soils. (The Japanese tailings investigated were fine sands and silts with cone resistances in the range from 0.2 to 6 MN/m^2 and initial voids ratios between 0.75 and 1.3.)

12.3   Hydraulic parameters

The theoretical relationships discussed below have not been proven in the field. They are put forward by way of illustration, and they are not intended to be used in practice.

12.3.1   Coefficient of consolidation

Two-dimensional, finite element analyses (Baligh and Levadoux, 1980; Acar et al., 1982) show that the rate of dissipation of excess pore pressures developed during cone penetration depends mainly on the horizontal permeability. Parametric studies reducing vertical permeability to one tenth of horizontal permeability showed little change in the rate of dissipation, although Acar et al. found a somewhat greater variation than did Baligh and Levadoux.

The theoretical solutions of Baligh and Levadoux and Acar et al. are shown in Figure 59(a) and (b), together with solutions based on cavity expansion theory by Torstensson (1977, 1982). The dissipation curves vary with differing positions of the pore-pressure filter. It can also be seen that there are considerable differences in the results of the analyses. Furthermore, Baligh and Levadoux emphasise that values of coefficient of horizontal consolidation, c_h, thus obtained are recompression values, and to evaluate coefficient of vertical consolidation, c_v, in virgin compression requires oedometer data.

The steps involved are as follows (May, 1983):

1. $c_{v\,(probe)} = c_{h\,(probe)} \cdot k_v/k_h$
   The ratio k_v/k_h is difficult to measure with any accuracy, but Ladd (1976) has given values of k_n/k_v for various soil types (Table 11).

2. $c_{v(NC)} = \left[\dfrac{RR}{CR} \right] c_{v\,(probe)}$
   where CR = compression ratio = $C_c/1 + e_o$
   RR = recompression ratio = $C_s/1 + e_o$

Tavenas et al. (1982) list many objections to the use of existing theoretical correlations. They point out that:

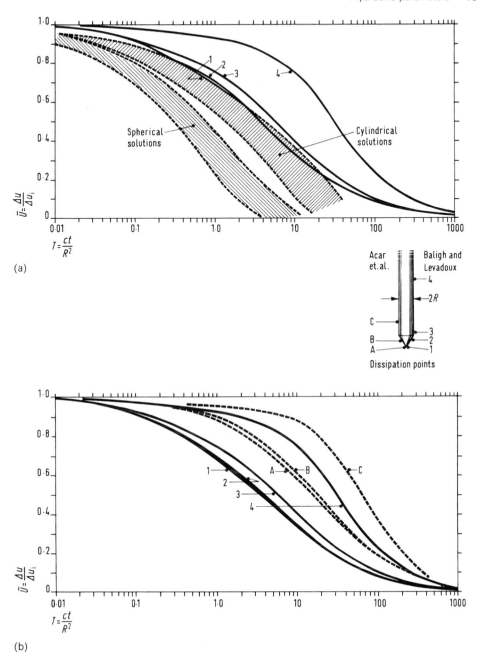

Figure 59 Pore-pressure dissipation plotted against time (a) Torstensson (1977, 1982) compared with Baligh and Levadoux (1980). (b) Acar, Tümay and Chan (1982) compared with Baligh and Levadoux (1980)

Table 11   Anisotropic permeability of clays (from Ladd, 1976)

Nature of clay	k_h/k_v
No evidence of layering	1.2 ± 0.2
Slight layering (e.g. sedimentary clays with occasional silt dustings to random lenses)	2 to 5
Varved clays in north eastern USA	10 ± 5

UK experience indicates considerably greater ratios of k_h to k_v in varved and layered deposits.

1. cavity expansion theories do not take account of the actual stress–strain behaviour
2. the correct shape for the expanding cavity cannot be modelled
3. the piezocone displaces (and thus remoulds) the clay in the immediate vicinity of the tip
4. rate effects are ignored.

In a NC soil, consolidation develops in a heterogenous system: the remoulded clay next to the probe has a low modulus and variable permeability during consolidation, and at an unknown distance the intact clay has a higher modulus and constant permeability. In addition, the pore-pressure dissipation around a piezocone initially corresponds to simultaneous consolidation and swelling, and laboratory tests show that compressibility and c_h differ in these two conditions.

In view of the shortcomings of theoretical correlations, Tavenas *et al.* prefer an empirical approach for the interpretation of pore-pressure dissipation. The approach they suggest is to correlate t_{50} (= time to 50% dissipation) with $k\,\sigma'_p/\gamma_w$, where the permeability, k, is obtained from *in-situ* permeability tests and the pre-consolidation pressure, σ'_p, is obtained from oedometer tests. The correlation they obtained for Champlain clays is

$$t_{50} = \frac{T_{50}r_o^2\gamma_w}{mk\sigma'_p}$$

where $T_{50} = \dfrac{1.2 \times 10^{-2}}{r_o^2}$

$m = M/\sigma'_p$

M = 'modulus of compressibility'

and σ'_p = preconsolidation pressure

The limitation of this approach is the need for laboratory work to determine σ'_p and m.

It can be seen that there are considerable difficulties involved at this stage in assessing c_h and c_v from pore-pressure dissipation. It is suggested that a rough value of c_h can be obtained by using the theoretical relationships of Baligh and Levadoux given in Figure 59. For a more accurate assessment, it is necessary to develop empirical correlations, for the particular clays concerned, by using *in-situ* permeability data and laboratory oedometer tests.

12.3.2   Permeability

Very approximate values of permeability, vertical and horizontal, can be obtained from:

$$k_v = c_v \cdot m_v \cdot \gamma_w$$
$$\text{and} \quad k_h = c_h \cdot m_h \cdot \gamma_w$$

The uncertainties in determination of c_h are apparent from the discussion in Section 12.3.1. As a first approximation, $m_h \simeq m_v$, so that $k_h \simeq c_h m_v \gamma_w$, and an estimate of m_v can be made as discussed in Section 6.4.1.

12.4   Synopsis: Engineering parameters from the piezocone

12.4.1   Undrained shear strength

Refer to Section 12.2.1.

Empirical relationships between pore pressure change and c_u have the form:

$$\Delta_u = N_{\Delta u} \cdot c_u$$

For some Canadian clays

$$N_{\Delta u} = 7.9 \pm 0.7, \text{ for } 0.8 < I_L < 2.0$$
$$N_{\Delta u} = 11.7 \pm 2.0, \text{ for } I_L > 2.0$$

For a soft, sensitive marine clay of OCR 2.1 to 2.3

$$N_{\Delta u} = 7.4$$
$$N_{\Delta u} = 4.7$$

Somewhat lower values would be expected in firm or stiff clays of low sensitivity.

Only limited data available on values of $N_{\Delta u}$ which depend on OCR. Tavenas *et al.* (1982). Δ_u measured at base of cone; c_u from *in-situ* vane tests.

Roy *et al.* (1982). For Δ_u measured at cone point

For Δ_u measured one diameter above the base of the cone. Local correlations are to be preferred.

12.4.2   Effective stress shear strength

Refer to Section 12.2.3.

Shear strength in effective stress terms may be expressed as

$$\tau_f = (\sigma' + a) \tan \phi'$$

where a is 'attraction' the negative intercept on a plot of $(q_T - \sigma_{vo})$ against σ'_{vo}.

Senneset and Janbu (1985).

Refer to Figure 58(c).

Use Figure 58(d) to determine $\tan \phi'$ with

$$N_m = \frac{q_T - \sigma_{vo}}{\sigma'_{vo} + a} = \frac{q_n}{\sigma'_{vo} + a}$$

and

$$B_q = \frac{u_T - u_0}{q_T - \sigma_{vo}} = \frac{\Delta_u}{q_n}$$

References

ACAR, Y. B., TUMAY, M. T. and CHAN, A. (1982) Interpretation of the dissipation of penetration pore pressures. Proc. Int. Symp. on Numerical Models in Geomechanics, Zurich, 1982. (Edited by R. Dungar, G. N. Pande and J. A. Studer). Balkema, Rotterdam, 1982

AMAR, S. (1974) The use of the static penetrometer in the Laboratoires des Ponts et Chaussées. Proc. 1st Eur. Symp. on Penetration Testing, Stockholm, 1974, Vol. 2.2, 7–12

AMAR, S., BAGUELIN, F. and JEZEQUEL, J. F. (1982) Pressio-penetrometer for geotechnical surveys. Proc. 2nd Eur. Symp. on Penetration Testing, Amsterdam, 1982, 419–423

AMERICAN SOCIETY FOR TESTING AND MATERIALS (1973) Evaluation of relative density and its role in geotechnical projects involving cohesionless soil. ASTM Special Technical Publication STP523, 1973

AMERICAN SOCIETY FOR TESTING AND MATERIALS (1975) Tentative method for deep quasi-static, cone and friction-cone penetration tests in soils. ASTM D3441–75T, 1975

ANAGNOSTOPOULOS, A. G. (1974) Evaluation of the undrained shear strength from static cone penetration tests in soft silty clay in Patras. Proc. 1st Eur. Symp. on Penetration Testing, Stockholm, 1974, Vol. 2.2, 13 and 14

BALDI, G. et al. (1981) Cone resistance of dry medium sand. Proc. 10th Int. Conf. on Soil Mechanics and Foundation Engineering, Stockholm, 1981, Vol. 2, 427–432

BALDI, G. et al. (1982) Design parameters for sands from CPT. Proc. 2nd Eur. Symp. on Penetration Testing, Amsterdam, 1982, 425–432

BALDI, G. et al. (1983) Static penetration tests and relative density of sand. 15th National Convention, Associazione Geotecnica Italiana, Spoleto, 1983

BALDI, G. et al. (1985) Penetration resistance and liquefaction resistance evaluation. Proc. Am. Soc. Civ. Engrs. – J. Geotech. Engng. Div. Dec. 1985 **111**(GT12), 1425–1445

BALIGH, M. M. and LEVADOUX, J. N. (1980) Pore pressure dissipation after cone penetration. Massachusetts Institute of Technology, Boston, Report MITSG 80–13, 1980

BALIGH, M. M. et al. (1981) The piezocone penetrometer (in Cone penetration testing and experience, Edited by G. M. Norris and R. D. Holtz). Proc. Session sponsored by Geotechnical Engineering Division of Am. Soc. Civ. Engrs., St Louis (Missouri), Oct. 1981, 247–263

BARENTSEN, P. (1936) Short description of a field testing method with a cone shaped sounding apparatus. Proc. 1st Int. Conf. on Soil Mechanics and Foundation Engineering, Harvard Univ., Boston (Mass), 1936, Vol. 2, 10

BATTAGLIO, M. et al. (1981) Piezometer probe tests in cohesive deposits (in Cone penetration testing and experience, Edited by G. M. Norris and R. D. Holtz). Proc. Session sponsored by Geotechnical Engineering Division of Am. Soc. Civ. Engrs. National Convention, St Louis (Missouri), Oct. 1981, 264–302

BATTAGLIO, M. and MANISCALCO, R. (1983) Il piezocono, esecuzione ed interpretazione. Atti dell' Istituto di Scienze delle Costruzioni de Politecnico di Torino, No. 607, 1983

De BEER, E. E. (1963) The scale effect on the transposition of the results of deep-sounding tests on the ultimate bearing capacity of piles and caisson foundations. Géotechnique Mar. 1963 **3**(No.1), 29–75

De BEER, E. E. (1977) Static sounding in clay and loam. Proc. Fugro 15-years Symp., Utrecht, 1977, Fugro BV, Leidschemdam (Netherlands), 1977, 43–68

De BEER, E. E. and MARTENS, A. (1957) Method of computation of an upper limit for the influence of the heterogeneity of sand layers on the settlement of bridges. Proc. 4th Int. Conf. on Soil Mechanics and Foundation Engineering, London, 1957, Vol. 1, 275–282

De BEER, E. E. et al. (1977) Bearing capacity of displacement piles in stiff fissured clays. Institute for Scientific Research in Industry and Agriculture, Brussels, Publication 39, 1977

BEGEMANN, H. K. (1953) Improved method of determining resistance to adhesion by sounding through a loose sleeve placed behind the cone. Proc. 3rd Int. Conf. on Soil Mechanics and Foundation Engineering, Zürich, 1953, Vol. 1, 213–217

BEGEMANN, H. K. (1965) The friction jacket cone as an aid in determining the soil profile. Proc. 6th Int. Conf. on Soil Mechanics and Foundation Engineering, Montreal, 1965, Vol. 1, 17–20

BEGEMANN, H. K. (1977) Soil mechanics aspects of pile foundations. Foundations Building Research SBR, Samson, Alphen a/d Rijn (Netherlands), 1977, 44–65

BEREZANTZEV, V. G., KHRISTOFOROV, V. and GOLUBKOV, V. (1961) Load bearing capacity and deformation of piled foundations. Proc. 5th Int. Conf. on Soil Mechanics and Foundation Engineering, Paris, 1961, Vol. 2, 11–15

van den BERG, A. P. (1982) Latest developments in cone penetrometers and other soil testing equipment. Proc. 2nd Eur. Symp. on Penetration Testing, Amsterdam, 1982, 447–455

BERINGEN, F. L., KOLK, H. J. and WINDLE, D. (1982) Cone penetration testing and laboratory testing in marine calcareous sediments. Proc. Symp. on Geotechnical Properties, Behaviour and Performance of Calcareous Soils (Edited by K. R. Demars and R. C. Cheney), Florida, Jan. 1981. American Society for Testing and Materials Special Technical Publication STP 777, 1982, 179–209

BERINGEN, F. L., WINDLE, D. and VAN HOOYDONK, W. R. (1979) Results of loading tests on driven piles in sand. Proc. Conf. on Recent Developments in the Design and Construction of Piles, 1979. Institution of Civil Engineers, 1980, 213–225

BJERRUM, L. (1972) Embankments on soft ground: state-of-the-art report. Proc. Conf. on Performance of Earth and Earth-supported Structures, Am. Soc. Civ. Engrs., Lafayette (Georgia), 1972

BRAND, E. W., MOH, Z-C. and WIROJANUGUD, P. (1974) Interpretation of Dutch cone tests in soft Bangkok Clay. Proc. 1st Eur. Symp. on Penetration Testing, Stockholm, 1974, Vol. 2.2 51–58

BRITISH STANDARDS INSTITUTION (1981) Code of practice for site investigations. BS 5930: 1981

BRUZZI, D. (1983) Underwater static penetrometer. Proc. Int. Symp. on Soil and Rock Investigations by In-situ Testing, Paris, 1983, Vol. 2, 223–226. International Association of Engineering Geology (c/o Laboratoire Central des Ponts et Chaussées, Paris)

BRUZZI, D. and CESTARI, F. (1982) An advanced static penetrometer. Proc. 2nd Eur. Symp. on Penetration Testing, Amsterdam, 1982, 479–486

BUILDING RESEARCH ESTABLISHMENT/NORWEGIAN GEOTECHNICAL INSTITUTE (BRE/NGI) (1985) Comparisons of piezocones in stiff overconsolidated clays. NGI, Oslo, Report 84223–1, 1985

BURBIDGE, M. C. (1982) A case study review of settlements on granular soils. MSc Dissertation, Imperial College of Science and Technology, London University, 1982

BURLAND, J. B. (1973) Shaft friction of piles in clays – a simple fundamental approach. Ground Engineering May 1973 6(No.3), 30–42

BURLAND, J. B. and BURBIDGE, M. C. (1985) Settlement of foundations on sand and gravel. Proc. Instn. Civ. Engrs. Dec. 1985 78(Part 1), 1325–1381

BURLAND, J. B. and LORD, J. A. (1969) The load deformation behaviour of Middle Chalk at Mundford, Norfolk: a comparison between full scale performance and in-situ and laboratory experiments. Proc. Conf. on In-situ Investigations in Soils and Rocks, British Geotechnical Society, London, 1969, 3–15

BURLAND, J. B., BROMS, B. B. and DE MELLO, V. F. B. (1977) Behaviour of foundations and structures: state-of-the-art review. Proc. 9th Int. Conf. on Soil Mechanics and Foundation Engineering, Tokyo, 1977, Vol. 3, 495–546

BUTLER, F. G. (1975) Heavily overconsolidated clays: state-of-the-art report, Proc. British Geotechnical Society Conf. on Settlement of Structures, Cambridge, 1974. Pentech Press (London), 1975, 531–610

BYRNE, P. M. and ELDRIDGE, T. L. (1982) A three parameter dilatant elastic stress-strain model for sand. University of British Columbia, Vancouver, Civil Engineering Department Soil Mechanics Series No. 57, May 1982

CAMPANELLA, R. G., GILLESPIE, D. and ROBERTSON, P. K. (1982) Pore pressures during cone penetration testing. Proc. 2nd Eur. Symp. on Penetration Testing, Amsterdam, 1982, 507–512

CAMPANELLA, R. G. and ROBERTSON, P. K. (1981) Applied cone research (in Cone penetration testing and experience, Edited by G. M. Norris and R. D. Holtz). Proc. Session sponsored by Geotechnical Engineering Division of Am. Soc. Civ. Engrs. National Convention, St. Louis (Missouri), Oct. 1981, 343–362

CAMPANELLA, R. G., ROBERTSON, P. K. and GILLESPIE, D. (1981) In-situ testing in saturated silt (drained or undrained?). Proc. 34th Canadian Geotechnical Conf., 1981

CANCELLI, A. (1983) Penetration tests on cohesive soils in Northern Italy, Proc. Int. Symp. on Soil and Rock Investigations by In-situ Testing, Paris, 1983, Vol. 2, 241–245. International Association of Engineering Geology (c/o Laboratoire Central des Ponts et Chaussées, Paris)

CARPENTIER, R. (1982) Relationship between cone resistance and the undrained shear strength of stiff fissured clays. Proc. 2nd Eur. Symp. on Penetration Testing, Amsterdam, 1982, 519–528

CHAPMAN, G. A. and DONALD, I. B. (1981) Interpretation of static penetration tests in sand. Proc. 10th Int. Conf. on Soil Mechanics and Foundation Engineering, Stockholm, 1981, Vol. 2, 455–458

CLARK, A. R. and WALKER, B. F. (1977) A proposed scheme for the classification and nomenclature for use in the engineering description of Middle Eastern sedimentary rocks. *Géotechnique* Mar. 1977 **27**(No. 1), 93–99

COUTTS, J. S. (1986) Correlations between piezocone results and laboratory data from six test locations. MSc Thesis, University of Surrey, Guildford, Feb. 1986

DAVIDSON, J. L. and BOGHRAT, A. (1983) Flat dilatometer testing in Florida. USA. Proc. Int. Symp. on Soil and Rock Investigations by In-situ Testing, Paris, 1983, Vol. 2, 251–255. International Association of Engineering Geology (c/o Laboratoire Central des Ponts et Chaussées)

DAVIS, R. O. and BERRILL, J. B. (1983) Comparison of a liquefaction theory with field observations. *Géotechnique* Dec. 1983 **33**(No.4), 455–460

DEMARS, K. R. and CHANEY, R. C. (Editors) (1982) Geotechnical properties, behaviour and performance of calcareous soils. Proc. Symp. sponsored by American Society of Civil Engineers, Florida, Jan. 1981. ASCE Special Technical Publication STP 777, 1982

DOUGLAS, B. J. and OLSEN, R. S. (1981) Soil classification using electric cone penetrometer (*in* Cone penetration and testing, Edited by G. M. Norris and R. D. Holtz). Proc. Session sponsored by Geotechnical Division at Am. Soc. Civ. Engrs. National Convention, St Louis (Missouri), Oct. 1981, 209–227

DUNCAN, M. J. and BUCHIGNANI, A. L. (1976) An engineering manual for settlement studies. University of California, Berkeley, 1976

DURGUNOGLU, H. T. and MITCHELL, J. K. (1975) Static penetration resistance of soils. Proc. Conf. on In-situ Measurement of Soil Properties, Am. Soc. Civ. Engrs. Raleigh (N. Carolina), June 1975, Vol. 1, 151–188

ENEL (MILAN), ISMES (BERGAMO) and UNIVERSITY OF TURIN (1985) Laboratory validation of in-situ tests (*in* Geotechnical Engineering in Italy, an overview 1985). Jubilee Volume for 11th Int. Conf. on Soil Mechanics and Foundation Engineering, San Francisco, 1985, 251–271

FRYDMAN, S. (1970) Discussion. *Géotechnique* Dec. 1970 **20**(No.4), 454 and 455

GEUZE, E. C. (1953) Resultats d'essais de penetration en profondeur et de mise en charge de pieux modeles. *Annales – Inst. Technique du Bâtiment et des Travaux Publics* 1953 **16**, 63 and 64

HANDY, R. L. *et al.* (1982) In-situ stress determination by Iowa stepped blade. *Proc. Am. Soc. Civ. Engrs – J. Geotech. Engng. Div.* Nov. 1982 **108**(GT11), 1405–1422

HARDIN, B. O. and DRNEVICH, V. P. (1972) Shear modulus and damping in soils: design equations and curves. *Proc. Am. Soc. Civ. Engrs – J. Soil Mech. Found. Div.* July 1972 **98**(SM7), 667–692

HEIJNEN, W. J. (1973) The Dutch cone test: study of the shape of the electrical cone. Proc. 8th Int. Conf. on Soil Mechanics and Foundation Engineering, Moscow, 1973, Vol. 1.1, 181–184

HEIJNEN, W. J. (1974) Penetration testing in Netherlands: state-of-the-art report. Proc. 1st Eur. Symp. on Penetration Testing, Stockholm, 1974, Vol. 1, 79–83

HENKEL, D. J. and WADE, N. H. (1966) Plane strain on a saturated remoulded clay. *Proc. Am. Soc. Civ. Engrs. – J. Soil Mech. Found. Div.* Nov. 1966 **92**(SM6), 67–80

HOBBS, N. B. and HEALY, P. R. (1979) Piling in chalk. CIRIA/DoE Piling Development Group Report PG6, Sept. 1979

JAMIOLKOWSKI, M. *et al.* (1982) Undrained strength from CPT. Proc. 2nd Eur. Symp. on Penetration Testing, Amsterdam, 1982, 599–606

JAMIOLKOWSKI, M. *et al.* (1985) New developments in field and laboratory testing of soils. Theme Lecture, 11th Int. Conf. on Soil Mechanics and Foundation Engineering, San Francisco, 1985

JANBU, N. and SENNESET, K. (1973) Field compressometer – principles and application. Proc. 8th Int. Conf. on Soil Mechanics and Foundation Engineering, Moscow, 1973, Vol. 1.1, 191–198

JANBU, N. and SENNESET, K. (1974) Effective stress interpretation of in-situ static penetration tests. Proc. 1st Eur. Symp. on Penetration Testing, Stockholm, 1974, Vol. 2.2, 181–193

JEFFERIES, M. G. and FUNEGARD, E. (1983) Cone penetration testing in the Beaufort Sea. Proc. Am. Soc. Civ. Engrs. Speciality Conf. on Geotechnical Practice in Offshore Engineering, Austin (Texas), 1983, 220–243

JONES, G. A. and RUST, E. (1982) Piezometer penetration testing CUPT. Proc. 2nd Eur. Symp. on Penetration Testing, Amsterdam, 1982, 607–614

JONES, G. A. and VAN ZYL, D. J. (1981) The piezometer probe – a useful investigation tool. Proc. 10th Int. Conf. on Soil Mechanics and Foundation Engineering, Stockholm, 1981, Vol. 2, 489–496

JONES, G. A., VAN ZYL, D. J. and RUST, E. (1981) Mine tailings characterisation by piezometer conc (in Cone penetration testing and experience. Proc. Session sponsored by Geotechnical Engineering Division of Am. Soc. Civ. Engrs., St Louis (Missouri), Oct. 1981, 303–324, Edited by G. M. Norris and R. D. Holtz). National Convention

JOUSTRA, K. and DE GIJT, J. G. (1982) Results and interpretation of cone penetration tests in soils of different mineralogic composition. Proc. 2nd Eur. Symp. on Penetration Testing, Amsterdam, 1982, 615–626

JUILLIE, Y. and SHERWOOD, D. E. (1983) Improvement of Sabkha soil of the Arabian Gulf coast. Proc. 8th Eur. Conf. on Soil Mechanics and Foundation Engineering, Helsinki, May 1983, 781–788

KJEKSTAD, O., LUNNE, T. and CLAUSEN, C. J. F. (1978) Comparison between in situ cone resistance and laboratory strength for overconsolidated North Sea clays. Marine Geotechnology 1978 3(No.1), 23–36. (also Norwegian Geotechnical Institute, Oslo, Publication 124, 1978)

KOMORNIK, A. (1974) Penetration testing in Israel: state-of-the-art report. Proc. 1st Eur. Symp. on Penetration Testing, Stockholm, 1974, Vol. 2.2, 245–252

LACASSE, S. and LUNNE, T. (1982) Penetration tests in two Norwegian clays. Proc. 2nd Eur. Symp. on Penetration Testing, Amsterdam, 1982, 661–670

LACASSE, S. et al. (1981) In-situ characteristics of two Norwegian clays. Proc. 10th Int. Conf. on Soil Mechanics and Foundation Engineering, Stockholm, 1981, Vol. 2, 507–512

LADANYI, B. (1967) Deep punching of sensitive clays. Proc. 3rd Panam. Conf. on Soil Mechanics and Foundation Engineering, Caracas, 1967, Vol. 1, 535–546

LADANYI, B. and EDEN, W. J. (1969) Use of the deep penetration test in sensitive clays. Proc. 7th Int. Conf. on Soil Mechanics and Foundation Engineering, Mexico, 1969, Vol. 1, 225–230

LADD, C. C. (1976) Use of precompression and vertical sand drains for stabilisation of foundation soils. Massachusetts Institute of Technology, Boston, Dept. of Civil Engineering, Report R76–4, 541, 1976

LADD, C. C. et al. (1977) Stress-deformation and strength characteristics. Proc. 9th Int. Conf. on Soil Mechanics and Foundation Engineering, Tokyo, 1977, Vol. 2, 421–424

LEDOUX, J. L., MENARD, J. and SOULARD, P. (1982) The penetro-gammadensimeter. Proc. 2nd Eur. Symp. on Penetrometer Testing, Amsterdam, 1982, 419–423

LEVADOUX, J. N. and BALIGH, M. M. (1980) Pore pressures during cone penetration. Massachusetts Institute of Technology, Boston, Dept. of Civil Engineering, Report R80–15, 1980

LUNNE, T. and CHRISTOFFERSEN, H. P. (1983) Interpretation of cone penetrometer data for offshore sands. Norwegian Geotechnical Institute, Oslo, Report 52108–15, 1983

LUNNE, T. and KLEVEN, A. (1981) Role of CPT in North Sea foundation engineering (in Cone penetration testing and experience, Edited by G. M. Norris and R. D. Holtz). Proc. Session sponsored by Geotechnical Engineering Division at Am. Soc. Civ. Engrs. National Convention, St. Louis (Missouri), Oct. 1981, 76–107

LUNNE, T., EIDE, O. and DE RUITER, J. (1976) Correlations between cone resistance and vane shear strength in some Scandinavian soft to medium stiff clays. Canadian Geotechnical J. Nov. 1976 13(No.4), 430 to 441. (also Norwegian Geotechnical Institute, Oslo, Publication 116, 1976)

MCCLELLAND, B. (1974) Design of deep penetration piles for ocean structures. Proc. Am. Soc. Civ. Engrs. – J. Geotech. Engng. Div. July 1974 100(GT7), 705–747

MARCHETTI, S. (1980) In-situ tests by flat dilatometer. Proc. Am. Soc. Civ. Engrs. – J. Geotech. Engng. Div. Mar. 1980 106(GT3), 299–321

MARCHETTI, S. (1982) Detection of liquefiable sand layers by means of quasi-static penetration tests. Proc. 2nd Eur. Symp. on Penetration Testing, Amsterdam, 1982, 689–695

MARR, L. S. (1981) Offshore application of the cone penetrometer (in Cone penetration testing and experience, Edited by G. M. Norris and R. D. Holtz). Proc. Session sponsored by Geotechnical Engineering Division of Am. Soc. Civ. Engrs. National Convention, St Louis (Missouri), Oct. 1981, 456–476

MARSLAND, A. (1971) Large in-situ tests to measure the properties of stiff fissured clays. Proc. 1st Australia/NZ Conf. on Geomechanics, Melbourne, 1971, Vol. 1. 180–189

MARSLAND, A. (1973) Laboratory and in-situ measurements of the deformation moduli of London Clay. Proc. Symp. on Interaction of Structure and Foundation, Midland Soc. of Soil Mechanics and Foundation Engineering, Univ. of Birmingham, 1971, 7 to 17. (also Building Research Establishment (Garston), Current Paper CP24/73, 1973)

MARSLAND, A. (1974) Comparison of the results from static penetration tests and large in-situ plate tests in London Clay. Proc. 1st Eur. Symp. on Penetration Testing, Stockholm, 1974, Vol. 2.2, 245–252

MARSLAND, A. (1975) In-situ and laboratory tests on glacial clays at Redcar. Proc. Symp. on Behaviour of Glacial Materials, Midland Soc. of Soil Mechanics and Foundation Engineering, Univ. of Birmingham, 1975, 164–180 (also Building Research Establishment (Garston), Current Paper 65/76, 1976)

MARSLAND, A. (1977) The evaluation of the engineering design parameters for glacial clays. *Q. J. Engng. Geol.* 1977 **10**(No. 1), 1–26

MARSLAND, A. (1980a) Discussion. Session 4, Proc. 7th Eur. Conf. Soil Mechanics and Foundation Engineering, Brighton, 1979, British Geotechnical Society (London), 1980.

MARSLAND, A. (1980b) The interpretation of in-situ tests in glacial clays. Proc. Int. Conf. on Offshore Site Investigation, Society for Underwater Technology, London, 1980, 218–228

MARSLAND, A. and QUARTERMAN, R. S. (1982) Factors affecting the measurements and interpretation of quasi-static penetration tests in clays. Proc. 2nd Eur. Symp. on Penetration Testing, Amsterdam, 1982, 697–702

MAY, R. E. (1983) The cone penetrometer with pore pressure measurement. University of Oxford, Dept. of Engineering Science, Report 1473/83, 1983

MAYNE, P. W. (1980) Cam clay predictions of undrained strength. *Proc. Am. Soc. Civ. Engrs. – J. Geotech. Engng. Div.* Nov. 1980 **106**(GT11), 1219–1242

MEIGH, A. C. (1969) Discussion. Proc. Conf. on In-situ Investigations in Soils and Rocks, British Geotechnical Society, London, 1969, 191 and 192

MEIGH, A. C. (1976) The Triassic rocks, with particular references to predicted and observed performance of some major foundations. 16th Rankine Lecture. *Géotechnique* Sept. 1976 **26**(No. 3), 391–452

MEIGH, A. C. and CORBETT, B. O. (1969) A comparison of in-situ measurements in a soft clay with laboratory tests and the settlement of oil tanks. Proc. Conf. on In-situ Investigations in Soils and Rocks, British Geotechnical Society, London, 1969, 173–179

MEIGH, A. C. and NIXON, I. K. (1961) Comparison of in-situ tests for granular soils. Proc. 5th Int. Conf. on Soil Mechanics and Foundation Engineering, Paris, 1961, Vol. 1, 499–507

MEYERHOF, G. G. (1951) The ultimate bearing capacity of foundations. *Géotechnique* Dec. 1951 **2**(No. 4), 301–332

MEYERHOF, G. G. (1956) Penetration tests and bearing capacity of cohesionless soils. *Proc. Am. Soc. Civ. Engrs. – J. Soil Mech. Found. Div.* Jan. 1956 **82**(SM1), 1–19

MEYERHOF, G. G. (1959) Compaction of sands and bearing capacity of piles. *Proc. Am. Soc. Civ. Engrs – J. Soil. Mech. Found. Div.* Dec. 1959 **85**(SM6), 1–30

MEYERHOF, G. G. (1974) State-of-the-art of penetration testing in countries outside Europe. Proc. 1st Eur. Symp. on Penetration Testing, Stockholm, 1974, Vol. 2.1, 40–48

MEYERHOF, G. G. (1976) Bearing capacity and settlement of pile foundations. *Proc. Am. Soc. Civ. Engrs. – J. Geotech. Engng. Div.* Mar. 1976 **102**(GT3), 195–228

MEYERHOF, G. G. (1983) Scale effects of ultimate pile capacity. *Proc. Am. Soc. Civ. Engrs. – J. Geotech. Engng. Div.* June 1983 **109**(GT6), 797–806

MUHS, H. and WEISS, K. (1971) Untersuchung von Grenztragfähigkeit und Setzungsverhalten flachgegrundeter Einzelfundamente in ungleichformigen nichtbindigen Boden (Investigation of ultimate bearing capacity and settlement behaviour of individual shallow footings in non-uniform cohesionless soil) (Summary in English). Degebo (Journal of the German Soil Mechanics Society), University of Berlin, 1971, Heft 26, 1–37

MUROMACHI, T. (1981) Cone penetration testing in Japan (*in* Cone penetration testing and experience, Edited by G. M. Norris and R. D. Holtz). Proc. session sponsored by Geotechnical Engineering Division at Am. Soc. Civ. Engrs. National Convention, St Louis (Missouri), Oct. 1981, 76–107

NASH, D. F. T. and DUFFIN, M. J. (1982) Site investigation of glacial soil using cone penetrometer tests. Proc. 2nd Eur. Symp. on Penetration Testing, Amsterdam, 1982, 733–738

NIEUWENHUIS, J. K. and SMITS, F. P. (1982) The development of a nuclear density probe in a cone penetrometer. Proc. 2nd Eur. Symp. on Penetration Testing, Amsterdam, 1982, 745–749

NIXON, I. K. (1982) Standard penetration test: state-of-the-art report. Proc. 2nd Eur. Symp. on Penetration Testing, Amsterdam, 1982, 2–24

NOORANY, I. and GIZIENSKY, S. F. (1970) Engineering properties of submarine soils: state-of-the-art review. *Proc. Am. Soc. Civ. Engrs. – J. Soil Mech. Found. Div.* Sept. 1970 **96**(SM5), 1735–1762

NORDLUND, R. L. (1963) Bearing capacity of piles in cohesionless soils. *Proc. Am. Soc. Civ. Engrs. – J. Soil Mech. Found. Div.* Mar. 1963 **89**(SM3), 1–35

NOTTINGHAM, L. C. (1975) Use of quasi-static friction cone penetrometer data to predict load capacity of displacement piles. PhD Dissertation, Dept. of Civ. Engng., University of Florida, Gainsville, 1975

O'RIORDAN, N. J., DAVIES, J. A. and DAUNCEY, P. C. (1982) The interpretation of static cone penetrometer tests in soft clays of low plasticity. Proc. 2nd Eur. Symp. on Penetration Testing, Amsterdam, 1982, 755–760

PARKIN, A. *et al.* (1980) Laboratory investigation of CPTs in sand. Norwegian Geotechnical Institute, Oslo, Report 52108–9, 1980

PHAN, T. N. (1972) Application of the Dutch cone test in Bangkok area. Asian Inst. Tech., Bangkok, MEng Thesis No. 377, 1972

POULOS, H. G. and DAVIS, E. H. (1980) Pile foundation analysis and design. John Wiley & Sons, London, 1980

POWER, P. T. (1982) The use of the electric cone penetrometer in the determination of the engineering properties of chalk. Proc. 2nd Eur. Symp. on Penetration Testing, Amsterdam, 1982, 769–774

RANDOLPH, M. F. and WROTH, C. P. (1979) An analytic solution for the consolidation around a driven pile. *Int. J. Numeric. Analytical Methods in Geomech.* 1979 **3** 217–229

RIDGEN, W. J. *et al.* (1982) A dual load range cone penetrometer. Proc. 2nd Eur. Symp. on Penetration testing, Amsterdam, 1982, 787–796

ROBERTSON, P. K. and CAMPANELLA, R. G. (1983) Interpretation of cone penetration tests: Parts 1 and 2. *Canadian Geotech. J.* Nov. 1983 **20** 718–745

ROBERTSON, P. K. and CAMPANELLA, R. G. (1985) Liquefaction potential of sands using the CPT. *Proc. Am. Soc. Civ. Engrs. – J. Geotech. Engng. Div.* Mar. 1985 **111**(GT3), 384–403

ROBERTSON, P. K., CAMPANELLA, R. G. and WIGHTMAN, A. (1982) SPT – CPT correlations. University of British Columbia, Vancouver, Civ. Engng. Dept., Soil Mechanics Series No. 62, 1982

ROL, A. H. (1982) Comparative study on cone resistance measured with three types of CPT tip. Proc. 2nd Eur. Symp. on Penetration Testing, Amsterdam, 1982, 813–819

ROY, M. *et al.* (1974) The interpretation of static cone penetration tests in sensitive clays. Proc. 1st Eur. Symp. on Penetration Testing, Stockholm, 1974, Vol. 2.2, 323–330

ROY, M. *et al.* (1981) Behaviour of a sensitive clay during pile driving. *Canadian Geotech. J.* 1981 **18**(No. 1), 67–85

ROY, M. *et al.* (1982) Development of pore pressures in quasi-static penetration tests in sensitive clay. *Canadian Geotech. J.* May 1982 **19**(No.2), 124–138

De RUITER J. (1975) The use of in-situ testing for North Sea soil studies. Proc. Offshore Europe Conf., Aberdeen, 1975, Paper OE–75219, Spearhead Publications, London, 1975

De RUITER, J. (1982) The static cone penetration test: state-of-the-art report. Proc. 2nd Eur. Symp. on Penetration Testing, Amsterdam, 1982, 389–405

SANGLERAT, G. (1979) The penetrometer and soil exploration. Elsevier Publishing Co., Amsterdam, 2nd Edition, 1979

SCHMERTMANN, J. H. (1969) Dutch friction-cone penetration exploration of research area at Field 5, Eglin Air Force Base, Florida. US Army Waterways Experimental Station, Vicksburg (Mississippi), Contract Report S–69–4, 1969

SCHMERTMANN, J. H. (1970) Static cone to compute settlement over sand. *Proc. Am. Soc. Civ. Engrs. – J. Soil Mech. Found. Div.* May 1970 **96**(SM3), 1011–1043

SCHMERTMANN, J. H. (1975) Measurement of in-situ shear strength. Proc. Conf. on In-situ Measurement of Soil Properties, Am. Soc. Civ. Engrs, Raleigh (N. Carolina), June 1975, 57–138

SCHMERTMANN, J. H. (1978) Guidelines for cone penetration test: performance and design. US Dept. of Transp., Fed. Highways Admin., Offices of Research and Development, Washington (DC), Report FHWA–TS–78–209, July 1978

SCHMERTMANN, J. H. (1981) Discussion: In-situ testing by flat dilatometer. *Proc. Am. Soc. Civ. Engrs. – J. Geotech. Engng. Div.* June 1981 **107**(GT6), 831–837

SCHMERTMANN, J. H., HARTMAN, J. P. and BROWN, P. R. (1978) Improved strain influence factor diagrams. *Proc. Am. Soc. Civ. Engrs. – J. Geotech. Engng. Div.* Aug. 1978 **104**(GT8), 1131–1135

SEARLE, I. W. (1979) The interpretation of Begemann friction jacket cone results to give soil types and design parameters. Proc. 7th Eur. Conf. on Soil Mechanics and Foundation Engineering, Brighton, 1979, Vol. 2, 265–270

SEED, H. B. and IDRISS, I. M. (1970) Soil moduli and damping factors for dynamic response analysis. Univ. of California, Berkeley, Report EERC 70–10, Dec. 1970

SEED, H. B. and IDRISS, I. M. (1971) Simplified procedure for evaluating soil liquefaction potential. *Proc Am. Soc. Civ. Engrs. – J. Soil Mech. Found. Div.* Sept. 1971 **97**(SM9), 1249–1273

SEED, H. B. and IDRISS, I. M. (1981) Evaluation of liquefaction potential of sand deposits based on observations of performance in previous earthquakes. Session on In-situ Testing to Evaluate Liquefaction Susceptibility, Am. Soc. Civ. Engrs. National Convention, St Louis (Missouri), Oct. 1981

SEED, H. B., TOKIMATSU, K. and HALDER, L. F. (1985) Influence of SPT procedures in soil liquefaction resistance evaluation. *Proc. Am. Soc. Civ. Engrs. – J. Geotech. Engng. Div.* Dec. 1985 **111**(GT12), 1425–1445

SEMPLE, R. W. and JOHNSON, J. W. (1979) Performance of 'Stingray' in soil sampling and in-situ testing. Int. Conf. on Offshore Site Investigation, Society for Underwater Technology, London, 1979, Paper 13, 169–181

SENNESET, K. and JANBU, N. (1985) Shear strength parameters obtained from static cone penetration tests.

Paper A–84–1, Inst. of Geotechnics and Found. Engng., Norwegian Inst. of Technology, Trondheim, Jan. 1985. (also presented at ASTM Symp., San Diego (California), 1984)

SEROTA, S. and LOWTHER, G. (1973) SPT practice meets critical review. *Ground Engineering* Jan. 1973 **6**(No. 1), 20–22

SHERWOOD, D. E. and CHILD, G. H. (1974) A static dynamic sounding technique. Proc. 1st Eur. Symp. on Penetration Testing, Stockholm, 1974, Vol. 2.2,7–12

SILVER, M. L. (1979) Automated data acquisition, transducers and recording for the geotechnical testing laboratory. *Geotechnical Testing J.* (ASTM) 1979 **2**(No.4), 185–189

SIMPSON, B. *et al.* (1979) Design parameters for stiff clays. General Report, 7th Eur. Conf. on Soil Mechanics and Foundation Engineering, Brighton, 1979, Vol. 5, 91–125

SKEMPTON, A. W. (1951) The bearing capacity of clays. Proc. Building Research Congress, Instn. Civ. Engrs. London, 1951, Vol. 1, 180–189

SKEMPTON, A. W. (1953) Discussion: piles and pile foundations, settlement of pile foundations. Proc. 3rd Int. Conf. on Soil Mechanics and Foundation Engineering, Zürich, 1953, Vol. 3, 172

SKEMPTON, A. W. (1957) Discussion on the planning and design on the new Hong Kong airport. *Proc. Instn. Civ. Engrs.* June 1957 **7** 305–307

SKEMPTON, A. W. and BJERRUM, L. (1957) A contribution to the settlement analysis of foundations on clay. *Géotechnique* Dec. 1957 **7**(No. 4), 168–178

SMITS, F. P. (1982) Penetration pore pressure measured with piezometer cones. Proc. 2nd Eur. Symp. on Penetration Testing, Amsterdam, 1982, 871–876

SUGAWARA, N. and CHIKARAISHI, M. (1982) On estimation of ϕ' for normally consolidated mine tailings using pore pressure cone penetrometers. Proc. 2nd Eur. Symp. on Penetration Testing, Amsterdam, 1982, 883–888

SWAIN, M. L. (1984) Developments in cone penetration testing – the friction piezocone. *Ground Engineering* Mar. 1984 **17**(No. 2), 26, 27

TAVENAS, S., LEROUEIL, S. and ROY, M. (1982) The piezocone test in clays: use and limitations. Proc. 2nd Eur. Symp. on Penetration Testing, Amsterdam, 1982, 889–894

te KAMP, W. C. (1977) Sonderen en funderingen op palen in zand (Static cone penetration testing and foundations on piles in sand). Fugro Sounding Symp., Utrecht, Oct. 1977

TERZAGHI, K. (1943) Theoretical soil mechanics. John Wiley & Sons, London, 1943

THORBURN, S. (1970) Discussion. Proc. Conf. on Behaviour of Piles, Instn. Civ. Engrs., London, 1970, 53, 54

THORBURN, S. (1982) The performance of a four-storey building founded on late glacial clays compared with CPT predictions. Proc. 2nd Eur. Symp. on Penetration Testing, Amsterdam, 1982, 895–903

THORBURN, S. and MacVICAR, R. S. (1970) Pile load tests to failure in the Clyde alluvium. Proc. Conf. on Behaviour of Piles, Instn. Civ. Engrs., London, 1970, 1–7

TORSTENSSON, B.-A. (1977) The pore pressure probe. Geotechnical Meeting, Norwegian Geotechnical Society, Oslo, 1977, Paper 34, 34.1–34.15

TORTENSSON, B.-A. (1979) The landslide at Tuve. Proc. Nordic Geotechnical Meeting, Helsinki, 1979, 557–572

TORTENSSON, B.-A. (1982) A combined pore pressure and point resistance probe. Proc. 2nd Eur. Symp. on Penetration Testing, Amsterdam, 1982, 903–908

TRINGALE, P. T. and MITCHELL, J. K. (1982) An acoustic cone penetrometer for site investigations. Proc. 2nd Eur. Symp. on Penetration Testing, Amsterdam, 1982, 909–914

TUMAY, M. T., ACAR, Y. and DESEZE, E. (1982) Soil exploration in soft clays with the quasi-static electric cone penetrometer. Proc. 2nd Eur. Symp. on Penetration Testing, Amsterdam, 1982, 915–921

TUMAY, M. T., BOGGESS, R. L. and ACAR, Y. (1981) Subsurface investigations with piezocone penetrometer (*in* Cone penetration testing and experience, Edited by G. M. Norris and R. D. Holtz). Proc. Session sponsored by Geotechnical Engineering Division at Am. Soc. Civ. Engrs., National Convention, St Louis (Missouri), Oct. 1981, 325–342

VEISMANIS, A. (1974) Laboratory investigation of electrical friction-cone penetrometer in sand. Proc. 1st Eur. Symp. on Penetration Testing, Stockholm, 1974, Vol. 2.2, 407–419

VERMEIDEN, J. (1948) Improved soundings apparatus, as developed in Holland since 1936. Proc. 2nd Int. Conf. on Soil Mechanics and Foundation Engineering, Amsterdam, 1948, Vol. 1, 280–287

VERMEIDEN, J. and LUBKING, P. (1948) Soil investigation. Proc. Symp. on Foundation Aspects of Coastal Structures, Delft, 1948, Vol. 2, V, 1a

VESIC, A. S. (1967) A study of bearing capacity of deep foundations: final report. School of Civil Engineering, Georgia Institute of Technology, Atlanta, 1967

VESIC, A. S. (1970) Tests on uninstrumented piles, Ogeechee River site. *Proc. Am. Soc. Civ. Engrs. – J. Soil Mech. Found. Div.* Mar. 1970 **96**(SM2), 561–584

VESIC, A. S. (1972) Expansion of cavities in infinite soil mass. *Proc. Am. Soc. Civ. Engrs. – J. Soil Mech. Found. Div.* March 1972 **98**(SM3), 265–290

VESIC, A. S. (1977) Design of pile foundations. National Co-operative Highways Research Program, Transportation Research Board, National Research Council, Washington (DC), Synthesis of Highway Practice 42, 1977

WAKELING, T. R. (1969) A comparison of the results of standard site investigation methods against the results of a detailed geotechnical investigation in the Middle Chalk at Mundford, Norfolk. Proc. Conf. on In-situ Investigations in Soils and Rocks, British Geotechnical Society, London, 1969, 206–212

WARD, W. H., BURLAND, J. B. and GALLOIS, R. M. (1968) Geotechnical assessment of a site at Mundford, Norfolk, for a large proton accelerator. *Géotechnique* Dec. 1968 **18**(No.4), 399–431

WEBB, D. L. (1969) Settlement of structures on deep alluvial sand sediments in Durban, South Africa. Proc. Conf. on In-situ Investigations in soils and Rocks, British Geotechnical Society, London, 1969, 181–188

WEBB, D. L., MIVAL, K. N. and ALLINSON, A. J. (1982) A comparison of the methods of determining settlements in estuarine sands from Dutch cone penetration tests. Proc. 2nd Eur. Symp. on Penetration Testing, Amsterdam, 1982, 945–950

WELTMAN, A. J. and HEALY, P. R. (1978) Piling in 'boulder clay' and other glacial tills. CIRIA/DoE Piling Development Group Report PG5, Nov. 1978

WIZZA, A. E., MARTIN, R. T. and GARLANGER, J. E. (1975) The piezometer probe. Proc. Conf. on In-situ Measurement of Soil Properties, Am. Soc. Civ. Engrs., Raleigh (N. Carolina), June 1975, Vol. 1, 536–545

WROTH, C. P. (1984) The interpretation of in-situ soil tests, 24th Rankine lecture. *Géotechnique* Dec. 1984 **34**(No. 4), 449–489

ZHOU, S. G. (1980) Evaluation of the liquefaction of sand by static cone penetration test. Proc. 7th World Conf. on Earthquake Engineering, Istanbul, 1980, Vol. 3, 156–162

ZHOU, S. G. (1981) Influence of fines on evaluating liquefaction of sand by CPT. Proc. Int. Conf. on Recent Advances in Geotechnical Engineering and Soil Dynamics, University of Missouri, Rolla, 1981, Vol. 1, 167–172

ZUIDBERG, H. M. (1975) Seacalf, a submersible cone penetrometer rig. *Marine Geotechnology* 1975 **1** 15–32

ZUIDBERG, H. M., SCHAAP, L. H. and BERINGEN, F. L. (1982) A penetrometer for simultaneously measuring cone resistance, sleeve friction, and dynamic pore pressure. Proc. 2nd Eur. Symp. on Penetration Testing, Amsterdam, 1982, 963–970

ZUIDBERG, H. M. and WINDLE, D. (1979) High capacity sampling using a drillstring anchor. Proc. Int. Conf. on Offshore Site Investigation, Society for Underwater Technology, London, 1979, Paper 11, 149–157

Appendix A
European manufacturers of CPT equipment

Note: Inclusion of a manufacturer's name in this list does not imply any approval by CIRIA of the company or of its equipment.

Trade Name	Manufacturer	Telephone	Telex	UK Agent
BAT	Bengt–Arne Torstensson AB Box 27194, S–102 52 Stockholm, SWEDEN	(46)* 8–630870	16584 BAT—S	
v.d. BERG	a.p.v.d. BERG BV Ijzerweg 4/PO Box 68 8440 AB Heerenveen, NETHERLANDS	(31) 5130– 31355	46229 VBERG—NL	Edeco (see below)
BORRO	Borros AB Box 3063, 17103, Solna, SWEDEN	(46) 8–272620	11669 BORRO—S	
GEOTECH	Geotech AB Datavägen 53, S–436 00 Askim, SWEDEN	(46) 431– 289920	27150 GEOTECH – S	Craelius Co. Ltd. Long March Daventry, Northants NN11 4DX Telephone 03272–3431 Telex 31600
GOUDA	Goudsche Machine- fabriek BV P.O. Box 125, 2800 AC Gouda NETHERLANDS	(31) 1820– 17055	20825 GMF–NL	ELE International Ltd. Eastman Way, Hemel Hempstead, Herts HP2 7HB Telephone 0442–50221 Telex 825239
ISMES	Istituto Sperimentalo Modelli e Strutture, s.p.a. viale Giulio Cesare 29 24100 Bergamo, ITALY	(39) 035– 243043	301249 ISMES — I	

MI PC	Laboratoire Central des	(33) 1–532	200361
	Ponts et Chaussées	3179	LCPARI–F
	58 bd Lefèbvre		
	75732 Paris Cedex 15		
	FRANCE		
EDECO	English Drilling	(44) 0422	51687
	Equipment Co. Ltd.	72843	EDECO–G
	Pilcon Division,		
	Linley Moor Road,		
	Huddersfield,		
	W. Yorks HD3 3RW		

---

* *International Dialing Code for that country*

Appendix B   Combined techniques and push-in devices

The thrust machine of a CPT rig can be used to push into the ground a variety of devices, including samplers, piezometers, dilatometers, pressure cells, and probes of the geophysical logging type. In addition, a penetrometer tip can incorporate other sensors to measure such parameters as inclination of the tip, temperature and acoustic response, or it can be combined with a pressuremeter or nuclear density probe. Some of these are discussed below.

1.   Combined techniques

1.1   Pressuremeter

A combined penetrometer and pressuremeter (the pressio-penetrometer) has been developed by Laboratoires des Ponts et Chaussées in Paris (Amar *et al.* 1982). It consists of an 89-mm dia. electric cone with a piezometer filter immediately above its base and surmounted by a pressuremeter module, also 89-mm dia. A friction sleeve can also be fitted. As presented by Amar *et al.*, the penetration is by means of a vibratory driver, but the authors also indicate that 'static' penetration can be used. The apparatus can be used on land or offshore.

1.2   Nuclear density probe

This device measures bulk density by a gamma ray back scatter technique, with a radioactive source near the point of the probe and a detector mounted a short distance above it. Ledoux *et al.* (1982) describe a prototype probe, 45 mm in diameter, which measures cone resistance and bulk density, for use in soft compressible soils. It is particularly useful in the detection of organic layers.

Nieuwenhuis and Smits (1982) describe a prototype cone penetrometer with density probe, 36 mm in diameter, for use in dense sands.

1.3   Conductivity probe

The Delft Soil Mechanics Laboratory have developed a soil conductivity probe (SCP) for the *in-situ* determination of the porosity of saturated sands. This consists of a penetrometer tip, with cone and friction sleeve, above which the tube is fitted with four electrodes set in insulation material. The tests are carried out in two stages. In the first stage, penetration of the SCP is halted at 0.2-m intervals of depth. An alternating current is applied to the two outer electrodes, and the electrical resistance of the soil is then measured across the two inner electrodes.

During the second stage, a water conductivity probe (WCP) is put down near to the SCP position. The WCP contains a measuring cell which is sucked full of water, at 0.2-m intervals of depth, and the resistivity of the water is determined. The porosity of the sand is determined from the ratio of the two measured resistivities. Because it is not possible to derive a theoretical relationship valid for all sands, it is necessary for a given sand to determine this in the laboratory by density calibration tests.

1.4   Other combinations

Marr (1981) describes the equipping of a penetrometer tip with a thermistor for cone resistance and temperature measurements in frozen and unfrozen soils in the Beaufort Sea. The thermistor was connected to the friction sleeve, which was inoperable in this application. Some time delay is required for stabilisation of temperatures.

Acoustic penetrometers, in which friction cone penetrometers were fitted with a microphone to monitor the acoustic response of the soil during cone penetration, are described by Muromachi (1981) and Tringale and Mitchell (1982). This application is in the development stage, but present indications are that the acoustic response may provide useful information on soil type and soil profile.

2   Push-in devices

2.1   Samplers

A piston sampler of the same outside diameter as the reference tip can be used to take small samples. It is not suited to extensive sampling, but it can be used for spot sampling where CPT results need clarification. The CPT rig can also be used to push in larger piston samplers.

Of greater interest is the Delft continuous sampler. This is available in two sizes: 29-mm dia. for visual examination and laboratory determination of density and index properties, and 66-mm dia. (Figure 60) for a wider range of laboratory tests. The smaller sampler can be pushed into the ground with a 50-kN or 100-kN rig, and the 66-mm dia. sampler (Figure 60) with a 200-kN rig. In soft clays or loose silty sands, samples up to 18 m in length can be obtained, or longer with a modified sampler.

The sampler is advanced in the ground by pushing on the outer tubes, the sample being automatically fed into an impregnated nylon stockinette sleeve. The sample, in its sleeve, passes into a thin-walled plastic inner tube containing a supporting fluid. The upper end of the nylon sleeve is attached to the top cap, which is connected by a tension cable to a fixed point above ground. The samples are cut into 1-m lengths, corresponding to the lengths of extension tube, and placed in sample cases, the 66-mm

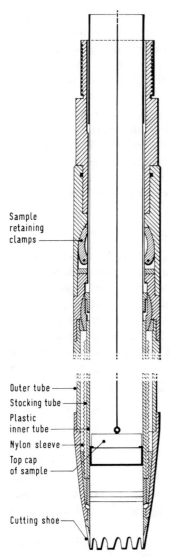

Sample
retaining
clamps

Outer tube
Stocking tube
Plastic
inner tube
Nylon sleeve
Top cap
of sample

Cutting shoe

Figure 60 Section through 66-mm Delft continuous sampler

samples being retained in plastic tubes. After specimens have been removed for testing, the remaining samples are split, then described and photographed when semi-dried. In this condition, the soil fabric is more easily identified. A 66-mm dia. stocking sampler for samples up to 2m in length, the Mostap, is produced by v.d. Berg (see Appendix A).

2.2 Dilatometers

The flat dilatometer (Marchetti, 1980) consists of a 95-mm wide stainless steel blade with a thin flat circular expandable membrane 60mm in diameter on one side (see

Figure 61). It is pushed into the ground with a penetrometer rig (or by other means) at a rate generally between 20 and 40 mm/s. Some information about ground conditions can be obtained during penetration, but the principal interest lies in the expansion stages which are undertaken at 200-mm depth intervals. During these, the membrane is expanded by means of gas under pressure, and the pressure required to move its centre 1.0 mm into the soil, within a time interval of between 15 and 30 s, is recorded. It is claimed that a 30-m profile can be obtained in about 2 h if no obstructions are met. From the results, quantitative estimates of K_o, OCR, m_v and c_u can be obtained on the basis of empirical correlations. Marchetti (1980) states that these can be used with a reasonable degree of confidence in a variety of insensitive soils, within certain specified limitations for each correlation.

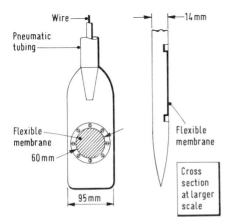

Figure 61 The Marchetti flat dilatometer (after Marchetti, 1980)

Davidson and Boghrat (1983) carried out a series of tests at five sites in Florida using the Marchetti dilatometer, CPT and a 'piezoblade', an electric piezometer probe identical in shape to the dilatometer. Disturbed and undisturbed samples were also obtained for comparative laboratory testing. The soils were of the Hawthorn Formation, consisting of varying amounts of clay (Montmorillonite and Kaolinite), quartz sand, and limestone with phosphatic grains and pebbles. General descriptions of the soils ranged from silty sand and sandy silt near the surface, through silt, clayey silt, and silty clay to clay at depth. For these Florida soils, the Marchetti correlations were found to overpredict OCR and constrained modulus and to underpredict the Young's modulus.

Lacasse and Lunne (1982) carried out flat dilatometer tests and piezocone tests in two soft Norwegian clays. They found that dilatometer test results in terms of K_o, OCR and M agreed approximately with values from field and laboratory tests, and the results adequately represented the soil variability with depth. The dilatometer produced values of compressibility which matched laboratory prediction within a factor of two or less. Schmertmann (1981) came to a similar conclusion.

Marchetti (1982) discusses the use of flat dilatometer testing in the evaluation of liquefaction potential. For this, use is made of the horizontal stress index, K_D, defined as

$$K_D = \frac{p_o - u_o}{\sigma_v'}$$

where u_o and σ_v' are the porewater pressure and vertical effective stress prior to blade insertion, and p_o is the pressure required to just begin to move the membrane (after correction for membrane stiffness).

It is suggested that K_D reflects a number of factors which influence liquefaction potential, although it is not possible to identify the individual effect of each factor. High values of K_D indicate favourable conditions. Low values of K_D indicate that the sand is loose, uncemented, in a low horizontal stress environment, and that it may be a source of liquefaction problems.

Experience with flat dilatometer testing (DMT) is limited at this stage, and further experience in various soils is required to confirm the empirical correlations. It is recommended that some DMT should be made on projects where CPTs are being carried out. In addition to furthering knowledge of DMT, the values of K_D so obtained may well assist in CPT interpretation.

In-situ stress measurements can also be made with the pressuremeter, or with the Iowa stepped blade (Handy *et al.*, 1982). This device compensates for installation disturbance by measuring soil pressures on three thicknesses of blade, in a step-tapered flat probe and extrapolating for the hypothetical pressure at zero blade thickness. *In-situ* soil stresses in fine-grained soils are found to agree with stresses from overburden pressures, elastic theory and pressuremeters. Measured K_o values in natural soil deposits are generally within acceptable ranges: 0.5 to 1.5 (depending on the consolidation state of the soil), and up to 4.2 in an expansive clay. Directional horizontal stresses can be obtained with a three-bladed stepped vane.

2.3 Piezometers

A cone penetrometer thrust machine provides an easy method of jacking piezometers into the ground. Where smear may occur, the piezometer should be of the type where the porous element is protected during penetration by a length of tube which is withdrawn to expose the element when the required depth is reached.

Appendix C
Specification check list*

	Refer to Appendix D
Thrust machine	
Max. thrust, and stroke	3.8
Mounting: trailer, truck, etc.; stability	
Reference tip	
Cone only or cone plus friction sleeve	3.1
Other sensors: inclination;	8.2
specials (See Appendix B)	
Load range: single or dual	
Tip geometry	3.1, 3.2
Manufacturing and operating tolerances	3.2
Gap and seal above cone	3.3
Friction sleeve	3.5, 6.7
Sensing devices (effect of load eccentricity)	3.4
Push rods	
Dimensions	3.6, 8.3
Straightness	3.6, 6.1
Guides	8.1
Friction reducer	3.9
Measuring equipment	3.7
Procedure	
Continuous testing: 20 ± 5 mm/s	4.1., 4.3, 4.4
Verticality	4.2, 8.2
Depth measurement	4.5
Precision of measurements	5

* The intention of this list is to give guidance on items for inclusion in a specification for cone penetration testing. The list is not comprehensive.

Checks and calibrations

see Table 12
(see Section 3.4)

Recording results

(See Section 3.3)
Information to be given on CPT plot   9.1
Information to be on file   9.2
Additional information   9.5
Scales for CPT plot   9.3
Site plan   9.4

Tips not conforming to reference test tip geometry   10.1–10.4

All divergences to be described explicitly   ⎰ 10.2.4
  ($B = ...,\ \alpha = ...,$ etc.)   ⎱ 10.3
Tolerances to be in the ratio
   $D/D_{\text{(standard)}}$ 10.2.3
Precision; two classes   10.6
Mechanical penetrometer:
  Correction for weight of inner rods ⎱
  Inner rods to slide easily   10.8.1.2
  and free from projections ⎰
  If manometers used (two)   7.1
  Discontinuous testing   10.4 and
                   11. (Notes
                   4 and 5)

Miscellaneous

Distance from borehole or previous CPT   6.3
Pre-threading of signal cable   6.5

Appendix D
Recommended standard for the cone penetration test

Appendix D (excluding Table 12) is reproduced from the 1977 Report of the ISSMFE Sub-committee on the PENETRATION TEST FOR USE IN EUROPE of which it forms Appendix A.

Table 12 Checks and calibration for Appendix D

Item	Ref to (App. D)	Frequency			Notes
		At start* of CPT programme	At Start of CPT profile	At end of CPT profile	
Verticality of thrust machine	4.2		●		
Straightness of push rods	3.6, 6.1		●		
Zero error				●	
Precision of measurements	5., 6.6	●			
Wear:		●			
dimension of cone	3.2, 6.2				
dimension of friction sleeve	3.5, 6.2				
roughness	3.5				
Seals:					
presence of soil particles	6.4		●		
quality	3.3, 6.4	●			
Calibration of load cells and proving ring	7.2	●			Or at 3-monthly intervals
Calibration of manometer	7.1	●			Or at 6-monthly intervals

* And regularly during a long programme

125

RECOMMENDED STANDARD FOR THE CONE PEN-
ETRATION TEST (CPT)

CONTENTS

1. SCOPE

The cone penetration test consists in pushing
into the soil, at a sufficiently slow rate,
a series of cylindrical rods with a cone at
the base, and measuring continuously or at
selected depth intervals the penetration resistance of the cone, and if required the
total penetration resistance and/or the friction resistance on a friction sleeve.

Cone penetration tests are performed in order
to obtain data on one or more of the following subjects:

1) the stratigraphy of the layers, and their
   homogeneity over the site

2) the depth to firm layers; the location of
   cavities, voids and other discontinuities

3) soil identification

4) mechanical soil characteristics

5) bearing capacity of piles

2. DEFINITIONS

2.1   CPT stands for Cone Penetration Test
and includes what has been variously called
Static Penetration Test, Quasi Static Penetration Test and Dutch Sounding Test.

2.2   Penetrometer (apparatus): an apparatus
consisting of a series of cylindrical rods
with a terminal body, called the penetrometer
tip and the measuring devices for the determination of the cone resistance, the local
side friction and/or the total resistance.

2.3   Penetrometer tip.

2.3.1   Penetrometer tip proper: the terminal
body at the end of the series of push rods,
which comprises the active elements that
sense the cone resistance, and the local side
friction resistance.

2.3.2   Conventional penetrometer tip: by
convention, if the length of the part of the
penetrometer tip proper located above the
cone is smaller than 1000 mm, the push rod
length to be added to the length of the penetrometer tip in order to obtain a length of
1000 mm.

2.4   Cone: the part of the penetrometer on
which the end bearing is developed.

According to the design of the apparatus the
following are distinguished:

2.4.1   Fixed cone penetrometer tip: the cone
can only be subjected to micro relative dis-

placements with respect to the other elements of the tip.

According to the shape of the cone the following are distinguished:

2.4.3   Simple cone in which the cylindrical prolongation above the conical part is generally equal to the diameter of the cone base.

2.4.4   Mantle cone: a cone which is prolonged with a more or less cylindrical sleeve, whose length is larger than the diameter of the base of the cone: this sleeve is called the mantle.

2.5   Friction sleeve: The section of the penetrometer tip upon which the local side friction to be measured is developed.

2.6   System of measurement.

The system includes the measuring devices themselves and the means of transmitting information from the tip to where it can be seen or recorded. For example the following can be distinguished:

2.6.1   Electric penetrometer: which uses electrical devices such as strain gauges, vibrating wires, etc..., built into the tip.

2.6.2   Mechanical penetrometer: which uses a set of inner rods to operate the penetrometer tip.

2.6.3   Hydraulic and pneumatic penetrometer: which uses hydraulic or pneumatic devices built into the tip.

2.7   Push rods: the thick walled tubes or rods used for advancing the penetrometer tip and, in addition, to guide and shield the measuring system.

2.8   Inner rods: solid rods which slide inside the push rods to extend the tip of a mechanical penetrometer.

2.9   Thrust machine: the equipment that pushes the penetrometer into the soil. The necessary reaction for this machine is obtained by dead weight or/and anchors.

2.10   Friction reducer: narrow local protuberance outside the push-rod surface, placed at a certain distance above the penetrometer tip, and provided to reduce the total friction on the push-rods.

2.11   Continuous and discontinuous penetration testing (see note 1 - para.11).

2.11.1   Continuous penetration testing: a penetration test in which the cone resistance is measured, while all elements of the penetrometer have about the same rate of penetration.

2.11.2   Discontinuous penetration testing: a penetration test in which the cone resistance is measured, while the other elements of the penetrometer tip remain stationary. When a friction sleeve is also included the sum of the cone resistance and resistance on the sleeve is measured when both cone and friction sleeve are pushed down, while the other elements of the penetrometer tip remain stationary.

2.12   The cone resistance q_c.

The cone resistance is obtained by dividing the total force acting on the cone Q_c, by the area of the base of the cone A_c

$$q_c = Q_c : A_c$$

This resistance is expressed in Pa, kPa or MPa [x]).

2.13   The local side friction f_s: the local unit side friction is obtained by dividing the force Q_s, needed to push down the friction sleeve, by its surface area A_s

$$f_s = Q_s : A_s$$

The local resistance f_s is expressed in Pa, kPa or MPa [x]).

2.14   Total force Q_t: the force needed to push cone and push-rods together into the soil. Q_t is expressed in kN.

2.15   Total side friction Q_{st}: this is generally obtained by subtracting the total force on the cone Q_c, from the total force Q_t

$$Q_{st} = Q_t - Q_c$$

Q_{st} is expressed in kN, as are Q_t and Q_c. Certain penetrometers allow Q_{st} to be measured directly.

2.16   Friction Ratio R_f and Friction Index I_f (see note 2 - para. 11).

2.16.1   Friction Ratio R_f: the ratio of the local side friction f_s to the cone resistance q_c, measured at the same depth, expressed as a percentage.

2.16.2   Friction index I_f: the ratio of the cone resistance q_c to the local side friction f_s, measured at the same depth.

---

[x]) 1 Pa (Pascal) = 1 N/m^2

3.   RECOMMENDED STANDARD PENETROMETER

3.1   General geometry of the penetrometer tip.

In the recommended standard penetrometer testing, penetrometer tips with or without a friction sleeve can be used (Fig. 1a and Fig. 1b).

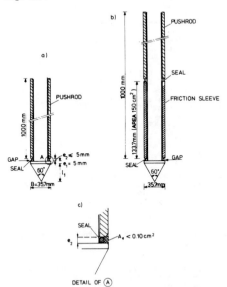

Fig 1.   Recommended standard penetrometer with a fixed cone and without (a) or with (b) a friction sleeve.

The penetrometer tip must have the same diameter as the cone over a length of 1000 mm above the cone base. The gap between the cone and the other elements of the penetrometer tip should be kept to the minimum necessary for the operation of the sensing devices, and designed and constructed in such a way as to prevent the entry of particles. This is also to apply to the gaps at either end of the friction sleeve, if one is included, and the other elements of the penetrometer tip. The axes of the cone, the friction sleeve if included and the body of the penetrometer tip, must be coincident.

In the case of a penetrometer tip without a sleeve, its diameter shall be the same as that of the base of the cone with a tolerance of -0.3 mm and +1 mm, over a length of 1000 mm (≈ 30 times the diameter of the base).

In the case of a penetrometer tip with a friction sleeve, the part of the penetrometer tip located above the friction sleeve shall have the same diameter as the friction sleeve over a length of 450 mm (≈ 12 times the diameter of the base) with a tolerance of -0.3 mm. The other parts of the penetrometer tip must also correspond with the above conditions for a penetrometer tip without a sleeve.

3.2   The cone.

The diameter of the base of the cone is 35.7 mm. The apex angle of the cone is $60°$.

The cone is to be continued by a cylindrical extension (Fig. 2); the height e_1 of this extension is 5 mm.

Manufacturing tolerances

on the diameter of the base of the cone	+0.3 mm
35.7 mm$\leq$B$<$36.0 mm	
on the height of the cone	+0.3 mm
31.0 mm$\leq$l$_1$$<$31.3 mm	
roughness of the cone < 5 μm.	

Operating tolerances

wear on the diameter of the base of the cone	-1 mm
34.7 mm$\leq$B$<$36.0 mm	
wear on the height of the cone	-7 mm
24.0 mm$\leq$l$_1$$<$31.3 mm	
wear on the length of the cylindrical extension	-2 mm

cones with a visible asymmetrical wear are to be rejected.

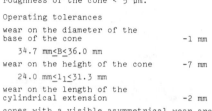

Fig 2.   Manufacturing (a) and operating (b) tolerances of the recommended standard cone.

3.3   Gap and seal above the cone (Fig. 1).

The gap between the cone and the other elements of the penetrometer must not be larger than 5 mm.

The seal placed in the gap should be properly designed and manufactured in order to prevent the entry of soil particles into the penetrometer tip. It must have a deformability many times larger than that of the sensing devices. The crosssectional area A_e of the gap, remaining after deduction of the area occupied by the seal must be smaller than 0.10 cm² (see cross-hatched area in Fig. 1c). The outer limits of the gap are to be shaped in such a way that the measurements are not affected by the possible entry of soil particles.

3.4   Sensing devices.

The sensing device should be designed to measure the cone resistance without being influenced by a possible eccentricity of that resistance.

3.5   Friction sleeve (Fig. 1b).

The diameter of the friction sleeve is to be manufactured and the sleeve retained in operation, only so long as it is at least the same value as the base of the cone, with a tolerance of +0.35 mm.

The surface area of the friction sleeve shall be 150 cm² with a tolerance of ±2%. The surface of the friction sleeve shall have a manufactured surface roughness of 0.5 μm (see note 3 para. 11) with a tolerance of ±50% in the direction of its longitudinal axis. In operation this roughness shall not become smaller than 0.25 μm. The projection above and below the friction sleeve shall correspond with the other parts of the penetrometer tip.

The friction sleeve is to be located immediately above the cone (Fig. 1b). The annular spaces between the friction sleeve and the other part of the penetrometer tip and their seals must conform to the same specifications as described under 3.3.

3.6   Push rods.

The push rods are screwed or attached together to bear against each other and to form a rigid-jointed series with a continuous straight axis. The deviation from the axis should not exceed 4‰ x) for the five lower push rods of the series and 8‰ x) for the remainder. The manner in which the "deviation" is determined is shown in Fig. 3.

When measuring the total friction with pushrods their diameter over the total length shall be 36 mm with a tolerance of ±1 mm.

---

x) These deviations corresponds in case of an even curvature to a deflexion of 1-2 mm in length.

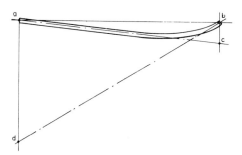

$$\frac{bc+ad}{ab} \leq 4‰ \quad \text{FOR THE 5 LOWER PUSHRODS}$$
$$\leq 8‰ \quad \text{FOR THE OTHER PUSHRODS}$$

Fig 3.   Determination of the deviation from the straight axis for push-rods.

3.7   Measuring equipment.

The resistances are to be measured by devices attached to the cone and the friction sleeve if included, and the signals are to be transmitted by a suitable method to a data recording system.

Exclusive recording of test results on a tape, which does not permit direct accessibility to the data, is not recommended.

3.8   Thrust machine.

The machine shall be able to provide a stroke preferably of one meter, and shall push the rods into the soil at a constant rate of penetration. The thrust machine shall be anchored and/or ballasted such that it does not move relative to the soil surface during the pushing action.

3.9   Friction reducer.

If a friction reducer is included, it should be located at least 1000 mm above the base of the cone.

4.   STANDARD TESTING PROCEDURE

4.1   Continuous testing.

The standard testing procedure is that of continuous penetration testing, in which the measurements are made while all elements of the penetrometer have the same rate of penetration.

4.2   Verticality.

The thrust machine is to be erected to obtain a thrust direction as near vertical as practicable. The maximum acceptable deviation of the thrust direction from the vertical is 2%. The axis of the push-rods must coincide with the thrust direction.

4.3   Rate of penetration.

The rate of penetration is the rate of the downward movement of the element of the penetrometer under consideration at the time the force on that element is measured.

The rate of penetration is 2 cm/sec with a tolerance of $\pm$ 0.5 cm/sec. This rate must be maintained during the entire stroke, even if readings are only taken at intervals.

4.4   Interval of readings.

A continuous reading is recommended. In no case shall the interval between the readings be more than 20 cm.

4.5   Measurement of the depth.

The depths are to be measured with an accuracy of at least 10 cm.

5.   PRECISION OF THE MEASUREMENTS

Taking into account all possible sources of error (parasitical frictions, errors of the recording devices, eccentricity of the load on the cone resp. the sleeve, temperature differences, etc...) the precision to be obtained should not be worse, than the larger of the following values:

    5% of the measured value
    1% of the maximum value of the range.

The precision must be verified in the laboratory or in the field taking into account all possible disturbing influences.

6.   PRECAUTIONS, CHECKS AND VERIFICATIONS

6.1   Before the CPT is made, the straightness of the push-rods, particularly of the lower five rods of the series, has to be checked. A method of checking the straightness consists in standing the push-rod vertically, spinning it, and observing whether it wobbles while it is rotating. If the wobble is noticeable, the push-rod should be discarded.

6.2   Regular inspections are to be made for wear (of the cone and friction sleeve).

6.3   It is also necessary to check that the

CPT test is not performed too close to existing boreholes or other penetrometer tests. For CPT tests with extended penetration it is recommended not to go closer than 25 boring diameters from uncased and unfilled boreholes, or at least 1 m from previously performed CPT tests.

6.4   The seals between the different elements of a penetrometer tip are to be regularly inspected to determine their quality. Prior to use the seals are to be checked to determine if soil particles are present.

6.5   Where the signals of the measuring devices built into the penetrometer tip are transmitted to the surface by an electric cable, it should be continuous, and consequently prethreaded through the push-rods.

6.6   Electric penetrometer tips should be temperature compensated. If the shift observed after extracting the tip is so large that the conditions of accuracy as defined under para. 5 are no longer met, the test should be discarded.

6.7   The friction sleeve transducer must operate in such a way that only shear stresses, and not normal stresses, are recorded.

7.   CALIBRATION

7.1   When manometers are being used, they are to be recalibrated at least every 6 months.

For each type of manometer there must be two identical units, each with its own calibration, available with the machine. At regular intervals the manometer used in the tests shall be checked against the reserve manometer.

7.2   The calibration of load cells or proving rings should be verified at least every 3 months.

Regular checks on the site with an appropriate field control unit are recommended.

8.   SPECIAL FEATURES

8.1   Push-rod guides.

Guides should be provided for the part of the push-rods protruding above the soil and for the push-rod length in water in order to prevent buckling.

8.2   Inclinometers.

In order to obtain more precise information

on the drift of the push-rods into the soil, inclinometers may be built into the penetrometer tip.

The need of such information depends on the soil conditions and increases with increasing depth of the test.

8.3   Push rods with smaller diameters.

In order to decrease the skin friction on the rods, use can be made of push-rods with a smaller diameter than that of the penetrometer tip. The distance between the smaller diameter push-rods and the cone base should be at least 1000 mm.

9.   REPORTING OF RESULTS

9.1   The following information shall be reported on the graphs of the measurements:

1. In order to state that the penetrometer and the test procedure are completely in agreement with the recommended Standard, each graph shall be marked with the letter S.
   After this letter will be added one of the following letters indicating the system of measurement:

   M = mechanical
   E = electric
   H = hydraulic

2. The date of the test and the name of the firm.

3. The identification number of the CPT test and the location of the site.

4. The depth from which a friction reducer, or push-rods with a reduced diameter, have been used. The depth at which the push-rods have been extracted over a limited height in order to break the lateral resistance, after which the push-rods have again been pushed into the soil.

5. Any abnormal interruption from the normal procedure of the CPT test.

6. Observations made by the operator such as soil type, sounds from the extension rods, indications of stones, disturbances, etc.

7. Data concerning the existence and thickness of fill, or existence and depth of an excavation, and level of the CPT test with respect to the original or artificial soil surface.

9.2   Besides the information indicated in para. 9.1, the internal files should also mention:

1. The identification number of the penetrometer tip used.

2. The name of the operator in charge of the crew which performed the test.

3. The dates and reference numbers of the calibration certificates for the measuring devices.

9.3   The following scales are recommended for the presentation of the graphs:

Depth scale: 1 unit length (arbitrary) for 1 m
Cone resistance q_c: the same unit length for 2 MPa
Local side friction f_s: the same unit length for 0.5 MPa
Total penetration force Q_t: the same unit length for 5 kN
Total friction Q_{st}:   ditto

So long as the above mentioned relationships between the scales along the vertical and horizontal axis are respected, the scales can be chosen arbitrarily in such a way that standard sized sheets can be used.

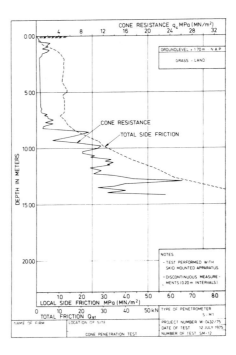

Fig. 4   An example of the presentation of test results from a CPT test.

9.4   Site plan.

For every investigation which is carried out, a clear site plan shall be drawn, with clear reference points in order that the locations of the penetrometer tests are accurately plotted.

Also when made in conjunction with borings the time sequences are to be indicated of the performance of the borings and CPT tests.

9.5   Besides the information mentioned under para. 9.1, it is recommended that the elevation of the soil-surface at the location of the test is given. In addition, where appropriate the following information should be given:

(a) The readings of the inclinometer, if taken.

(b) All checks made after extracting the push-rods, the conditions of the push-rods and the penetrometer tips.

(c) The depth of the water in the hole remaining after withdrawal of the penetrometer, or the depth at which the hole collapsed.

(d) Whether or not the testhole has been backfilled, and if so by which method.

10.   DIVERGENCES FROM THE STANDARD RECOMMENDED

10.1 General.

A general and very important specification is, that all divergences from the Recommended Standard are to be described explicitly and completely on the test graphs. In order to simplify these remarks the names or symbols defined in para. 10.5 can be used.

10.2 Divergences only related to the dimensions and the shape of the cone.

10.2.1 Diameter of the base of the cone.

In order to be able to penetrate deeper in certain cases a cone with a smaller base can be used. In order to be able to include the measuring device, or to be able to drive the penetrometer tip through hard layers with less danger of damage occurring to the tip, cones with larger diameters are used.

10.2.2 Apex angle of the cone.

To decrease the possibility of damage, an apex angle of 90° can be used.

10.2.3 Tolerances.

All tolerances specified for the Recommended Standard, are to be adapted in direct proportion to the diameter.

10.2.4 Symbols.

Tests performed with a diverging cone cannot be represented by the letter S, as they differ from the Standard. If all other elements are identical as in the Standard, they can be indicated by the letters M, E, H, but followed by the indication B =   and α = giving the values of the diverging characteristics of the cone.

10.3 Divergences only related to the location or dimensions of the friction sleeve.

10.3.1 If the friction sleeve, contrarily to the Recommended Standard, is not placed immediately above the base of the cone, the minimum distance (h) between that base, and the lower end of the friction sleeve should be three times the diameter of the base.

10.3.2 Surface of the sleeve.

When using a cone having a base diameter of 35.7 mm, but with a friction sleeve of a length other than the Recommended Standard, then the surface area of the sleeve should not be larger than 350 cm², and not smaller than 100 cm².

When using a cone with a base diameter different from the Recommended Standard, the surface area of the sleeve should be adjusted proportionally to the surface area of the base of the cone.

10.3.3 Symbols.

Tests performed with a diverging friction sleeve cannot be represented by the letter S, as they differ from the Recommended Standard. If all other elements are identical as in the Recommended Standard, they can be indicated by the letters M, E, H, but followed by the indication:

> height of lower end of sleeve h =
> surface area of the sleeve A_S =

10.4 Discontinuous testing with freee cone penetrometer tips.

10.4.1 With a free cone penetrometer tip, either continuous or discontinuous testing is possible. The manner in which the test is performed should be described in the report and on the test graphs.

10.4.2 Discontinuous testing.

In the case of discontinuous testing, although the rate of downward movement due to the thrust machine is known, the rate of penetration of the free cone at the point of rupture of the soil can be different to that of the movement due to the thrust machine. They only correspond when there is continuous downward movement of the push-rods.

When testing discontinuously the minimum

movement to be imposed on the cone or on the
friction sleeve is 0.5 times the diameter of
the cone (see note 4 para. 11).

10.5 Table of traditional penetrometer tips
diverging from the Recommended Standard.

The penetrometer tips actually in use in
several countries, and diverging from the
Recommended Standard, are given below. They
are indicated by a name, and a symbol which
has been added to permit abbrevations when
referring to them on the test graphs.

10.5.1 Mechanical penetrometer tip - note 5
para. 11.

M1   The Dutch mantle cone penetrometer
        tip (Fig. 5)
M2   The Dutch friction sleeve penetrom-
        eter tip (Fig. 6)
M3   The U.S.S.R. mantle cone penetrom-
        eter tip (Fig. 7)
M4   The simple cone penetrometer tip
        (Fig. 8)
M5.1   The Andina cone penetrometer tip
        (Fig. 9a)
M5.2   The Andina friction sleeve cone pen-
        etrometer tip (Fig. 9b)

10.5.2 Electric cone penetrometer tip.

E1.1   The Delft electric penetrometer tip
        (Fig. 10a)
E1.2   The Delft friction sleeve electric
        penetrometer tip (Fig. 10b)
E2   The Degebo friction sleeve electric
        penetrometer tip (Fig. 11)

10.5.3 Hydraulic penetrometer tip.

H1.1   The Parez hydraulic penetrometer tip
        (Fig. 12a)
H1.2   The Parez friction sleeve hydraulic
        penetrometer tip (Fig. 12b)

10.6 Precision of the measurements.

When testing diverges from the Recommended
Standard two classes of precision are defined:

        the normal precision class: see section
        5
        the lower precision class: the pre-
        cison obtained should not be worse than
        the larger of the following values:
            10% of the measured value
            2% of the maximum value of the
            range

In all such cases the class of precision of
the tests shall be indicated in the report
and on the test graphs.

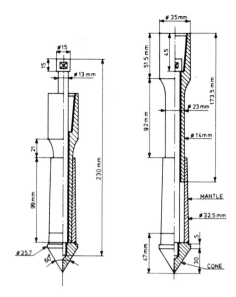

Fig. 5 The Dutch mantle cone penetrometer tip.
        Symbol: M1.

10.7 Static dynamic penetrometers and pre-
boring cone penetrometers.

Penetration can be increased by the use of
static dynamic penetrometers and also by the
use of penetrometers equipped with preboring
tools.

It must be clearly indicated in the report
and on the test graphs when such equipment
has been used.

10.8 Precautions, checks and verifications.

10.8.1 Mechanical penetrometers.

10.8.1.1 Push-rods.

There should not be any protruding edge on
the inside of the push-rods at the screw
connection between the rods (Fig. 13).

10.8.1.2 Inner rods.

The diameter of the inner rods shall be 0.5
to 1 mm less than the internal diameter of
the push-rods. The inner rods must slide very
easily through the push-rods.

The ends of the inner rods should be exactly
at right angles to the axis of the rod and
be machine-tooled to a smooth surface.

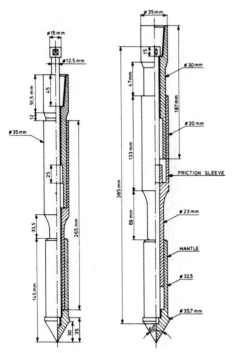

Fig. 6 The Dutch friction sleeve penetrometer tip.
Symbol: M2.

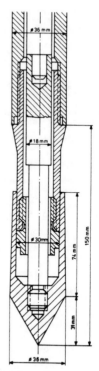

Fig. 7 The U.S.S.R. mantle cone penetrometer tip.
Symbol: M3.

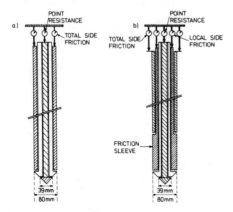

Fig. 9 The Andina cone penetrometer tip (a) and the
friction sleeve cone penetrometer tip (b).
Symbols: M5.1 and M5.2 respectively.

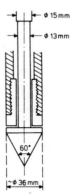

Fig. 8 The simple cone penetrometer tip.
Symbol: M4.

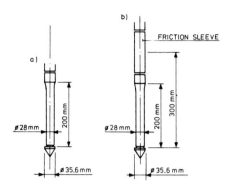

Fig. 10   The Delft electric penetrometer tip (a) and
the friction sleeve electric penetrometer
tip (b).
Symbols: E1.1 and E1.2 respectively.

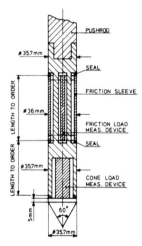

Fig. 11   The Degebo friction sleeve electric penetrometer tip.
Symbol: E2.

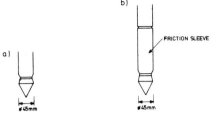

Fig. 12   The Parez hydraulic penetrometer tip (a) anu
the friction sleeve hydraulic penetrometer
tip (b).
Symbols: H1.1 and H1.2 respectively.

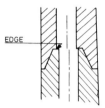

Fig. 13   There should not be any protruding edge at
the screw connections between the rods.

The rods must not screw together or be
joined in any way which gives them an additional degree of freedom as this has been
found to increase the parasitical friction
between the rods and the tubing. Before and
after the test a check will be made that the
inner rods slide very easily in the push-
rods, and that the cone and the friction
sleeve move easily with respect to the body
of the penetrometer tip. For improved accuracy at low values of the resistances, the
thrust data registered at the surface should
be corrected for the total weight of the
inner rods in the case of the cone resistance, and for that of the push-rods and inner
rods in the case of the total resistance.

11.   EXPLANATORY NOTES AND COMMENTS

Note 1: Definitions 2.11

The denominations "continuous" and "discontinuous" penetration testing are not quite
correct, and the denomination "penetration
testing with simultaneous pushing, and non-
simultaneous pushing of cone and push rods",
would be more adequate. However, the denominations "continuous" and "discontinuous"
have been maintained, as they are already
well established.

Note 2: Definitions 2.16

The friction ratio R_f, being the ratio of the

local side friction f_s to the cone resistance q_c, has to be expressed as a percentage in order to obtain a figure larger than one. Although in the past the friction ratio has been mostly used, there is a tendency to use the friction index I_f, being the ratio of the cone resistance q_c to the local side friction f_s, and which gives directly a figure larger than one. This is the reason why the two quantities have been included in the definitions.

Note 3: The friction sleeve 3.5

The roughness is defined as the mean deviation of the real surface of a body from the mean plane. The roughness is expressed in micrometers (μm).

Note 4: Discontinuous testing 10.4.2

In the case of mechanical penetrometer tips, in order to be certain that the cone and the friction sleeve move sufficiently with respect to the push-rods, due account is to be taken of the elastic shortening of the inner rods. Therefore, at the surface the movement of the inner rods relative to the push-rods must be at least equal to the sum of the minimum imposed movement of the cone plus the elastic shortening of the inner rods.

Note 5: Mechanical penetrometer tips 10.5.1

Continuous testing with a mechanical penetrometer tip is not recommended for high accuracy, as the movement of the inner rods to the push-rods can change its sense at different depths, increasing the margin of the error due to the parasitical internal friction. Furthermore it is necessary to check at least every meter during the test that the inner rods are still free to move relative to the push-rods.

Note 6: 10.5.1 M4

In the case of the simple cone special precautions have to be taken against soil entering the sliding mechanism and affecting the resistance. After extracting the penetrometer tip a check is to be made, in order to be certain that the cone-stem still moves completely freely relative to the bush.

Index